Energy Resources

The Author

Andrew L. Simon (Ph.D. Purdue University, 1962) is Professor and Head of the Civil Engineering Department at the University of Akron. A native of Hungary, Dr Simon has extensive engineering experience in the United States and abroad. His professional interests are in the fields of technology assessment and life style modification.

Energy Resources

ANDREW L. SIMON

Department of Civil Engineering
The University of Akron

PERGAMON PRESS INC.

New York · Toronto · Oxford · Sydney · Braunschweig

PERGAMON PRESS INC.
Maxwell House, Fairview Park, Elmsford, N.Y. 10523

PERGAMON OF CANADA LTD.
207 Queen's Quay West, Toronto 117, Ontario

PERGAMON PRESS LTD.
Headington Hill Hall, Oxford

PERGAMON PRESS (AUST.) PTY. LTD.
Rushcutters Bay, Sydney, N.S.W.

PERGAMON GmbH
Burgplatz 1, Braunschweig

Copyright © 1975, Pergamon Press Inc.

Library of Congress Cataloging in Publication Data

Simon, Andrew L
    Energy resources.

    Includes index.
    1. Power resources. I. Title.
TJ153.S587   1975   621   74-28320
ISBN 0-08-018750-1
ISBN 0-08-018751-X pbk.

Printed in the United States of America

To my sons, Anthony and Andrew

Contents

PREFACE   ix

Chapter 1 The Historical Perspective   1
        2 What is Energy?   9
        3 Ways We Use Energy   17
        4 Coal   33
        5 Oil   49
        6 Natural Gas   67
        7 The Heat Energy of the Underground   75
        8 Power from Sunshine   83
        9 The Energy of Running Waters   95
       10 To Catch the Wind   105
       11 Atomic Energy   111
       12 Fusion, the Promise of Limitless Power   125
       13 Other Techniques of Energy Conversion   131
       14 The Challenge   145

PROBLEMS TO PONDER   151
REFERENCES   157
INDEX   163

Preface

During the remainder of this century the procurement, utilization, and conservation of energy will be one of the central issues facing the people of America and the rest of the world. Until now we have been blessed with cheap and abundant energy. In the past the average citizen took it for granted that he could turn on the light by flipping a switch and he could have his car's tank filled by stopping at a gas station. How this came about was not his concern. It was up to the businessmen, economists, engineers, and scientists. Various specialists working in the field of energy production did their jobs quietly within their industrial, academic, or governmental organizations. Their interests were generally confined to the narrow range of their specialities. Geologists were searching for coal, oil, or uranium deposits. Physicists studied the complexities of nuclear fission. Hydraulic engineers built water power stations. Chemical engineers ran refineries, and mechanical engineers designed boilers, turbines, engines, and other energy converters. Businessmen, economists, even environmentalists contributed their knowledge toward producing the energy we needed to support our lifestyle.

The average American citizen had no reason to be concerned with the ways and means energy was obtained. At least not until something went wrong. Scarcities, rising costs, and the various effects of the 1973 energy crisis caused the American public to realize that the days of cheap and unlimited energy were over and the availability of power was not guaranteed by the Constitution. Suddenly, topics concerning the resources, utilization, and conservation of energy became everyday matters in the news media. Long lines of cars in front of the gas pump accomplished more overnight than hundreds of speeches and articles by thoughtful and concerned energy experts who for years had warned us about the growing gap between energy supply and demand. Today energy is an overwhelming public concern.

This book is written for the average concerned American who is rightfully but belatedly trying to find out what went wrong and why, what to do now, where he can help. In our democratic society it is imperative that the public be well informed about the fundamental issues facing the nation. It is the public that ultimately will make the decisions, bear the sacrifices, pay the costs, and enjoy the benefits. The public therefore must have the facts, brought forward from the pages of technical journals, engineering textbooks, and government reports and presented in plain, nontechnical language. This book is an attempt to do this.

In the following pages, first the past and then the present modes of energy production and utilization are presented along with explanations of what energy really is. Following these, the various energy resources are taken up one by one. Sources, availabilities, methods of procurement, future prospects, and current technical difficulties are discussed. Energy resources—coal, oil, natural gas, geothermal, solar, wind and water power, atomic fission and fusion—are described. After these chapters, new methods of energy conversion are surveyed that may have a significant impact on our power generation in the future. Finally some nontechnical considerations are presented. These are matters that may have to be taken into account as we attempt to overcome the technological difficulties in energy supply for the future.

No specific technical and scientific knowledge is needed to understand the subjects discussed. Admittedly, this limits the scientific depth of the text. The more demanding reader with scientific inclinations is advised to read the more specialized and advanced texts. There is no shortage of such advanced literature written by well-qualified authors.

It is my sincere hope that some of the readers of this book may be encouraged to enter a scientific or technical career in order to help provide for the future energy needs of the country. Others with nontechnical interests will perhaps benefit by learning how deeply and fundamentally their lives will be influenced in the future by the various subjects discussed in the following pages.

CHAPTER 1

The Historical Perspective

The twilight of civilization was perhaps marked by the first fire lit by a primeval caveman. Making fire was the first conscientious utilization of the world's energy resources. Before the use of fire, no doubt, there was intelligent life. Human life, like that of the animals, was made possible and was maintained by the sun. Perhaps the subconscious realization of this fact has given rise to religious veneration of the sun in most known early civilizations all over the world (65). The primordial father of gods before the first Egyptian dynasty at Heliopolis (Sun-city) was the sun god, Re. In Assyria and Babylonia, Gibil—the fire god—was the son of Anu—supreme god of the highest heaven. The god of light, from Heimdall of the Teutons to Atea of the Marquesas Islanders, was revered. Sun goddess Arinna of the Hittites in the Fertile Crescent and Amarasu of Japan were in a central position in the religious hierarchy. Sun gods of the most diverse nations include Svarog of the Slavs, Isten of the ancient Hungarians, Istanu of the old Turks, Inti of the Peruvian Indians, Zeus of the Greeks, Jupiter of the Romans, Varuna Mitra and Dyanus Pitar of the Indo Europeans. According to native legends in New Zealand, on the Marquesas and on the Chatham Islands, fire was borrowed from the gods. Slavonic rural folk still retain a mystic respect for fire—a sacred character. Their old forbid the young to swear or shout when fire is lighted in the house. Old Slavs (before Christianity, it is said) have prayed to fire, calling him Svarogich.

Prometheus the Titan, the originator and supporter of mankind according to ancient Greek legend, stole fire from heaven and gave it to man. A similar legend of the Sunka tribe of the South Seas gave this honor to a

1

man called Emakong. He brought fire from the land of the snakemen. In Kyoto, Japan, pure fire is made by friction of pieces of Hinoki wood (Kiri-Bi fire) or by priests at the temple of Gion striking stone with steel (Uchi-Bi fire). On New Year's Day the people carry this pure fire home to light their hearth, thereby receiving the god's protection throughout the year.

A less religious approach was taken by the ancient Greeks, who considered fire to be one of the four elements (together with Water, Earth, and Air) making up the world. Primitive man, before the use of fire, used energy only in the form of the food he ate. At best, he used about 2000 kilocalories* daily—if he found it. Hunting man had more food. The use of fire, for over 100,000 years, provided more energy. Man has used about 5000 kilocalories per day, some of it in the form of firewood. The first agricultural man (about 5000 B.C. in the Mediterranean area) used the sun's energy in growing crops and used animal energy as well. (Animal energy is also derived from the sun.) He probably used about 12,000 kilocalories. Use of animal oil (whales, etc.) by Eskimos, and naphtha by those who found it, is shrouded in antiquity. The Chinese drilled bore holes 3000 feet deep for natural gas as early as 1000 B.C. They used bamboo for piping and burned the gas for light and heat and for the evaporation of brine.

Primitive wheels harnessed water power in ancient Babylon as well as on the Nile in Egypt and on the Yellow River in China. Wind was used to move ships thousands of years before Christ, although manpower in galleons was the prime mover for about a thousand years thereafter.

The advanced agricultural man (through the Middle Ages of our civilization) used about 7000 kilocalories as food, 12000 in his home and commerce, 7000 in his agriculture and primitive industry, and 1000 for transportation—a total of 27,000 (31) per day.

In contrast to these figures, by 1870 the daily consumption of energy reached 70,000 kilocalories per capita in Western Europe and the United States. In 1970, this figure rose to about 230,000 kilocalories per capita per day in the United States (19). This data is depicted in Fig. 1.1.

The sophistication of energy converters paralleled man's development. The flint stones of the caveman concentrated his power. The lever and the block and tackle magnified man's force. The wheel overcame friction losses while dragging heavy objects. Man's technical knowledge increased rapidly during the Middle Ages. Medieval Europe saw the

*For the definition of kilocalorie, see Chapter 2.

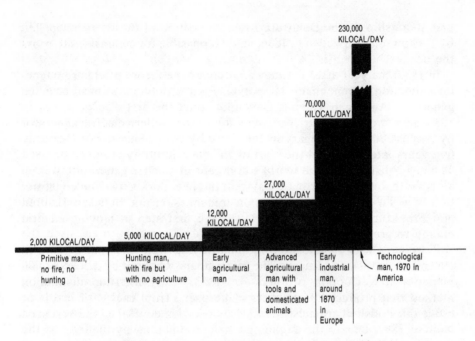

Fig. 1.1 Daily energy use by individuals through human history.

development of water power to harness the power of creeks, rivers, and even the tides of the Atlantic. The development of modern hydraulic turbines dates back to 1750, when J. A. Segner built a mill driven by an impulse turbine. J. B. Francis, an American, made significant progress in hydraulic turbine design in 1849.

Coal was introduced for heating and metallurgy in England during the 12th century and it spread rapidly to the continent. The need for pumping large amounts of water from deep mines provided the impetus for research. By 1698, Thomas Savery developed the first steam-driven pump. In 1712 Thomas Newcomen created a steam engine to drive pumps. Theoretical and practical progress went hand in hand. Joseph Black, a Scottish chemist, explained the theoretical concepts of steam and heat energy in 1764. His friend, James Watt, used these concepts in 1769 to develop his rotary steam engine, which was to revolutionize technology. By 1802, Richard Trevithick built a locomotive. This was improved by George Stephenson in 1814, who used it to transport coal from the mines. Soon these developments started the phenomenal expansion of rail transportation. In 1787 John Fitch of Connecticut built and operated the first steamship on the Delaware River. Invention of the screw propeller, by J. Ericsson of Sweden in 1836, made it possible for steamships to cross the oceans.

In 1832, the year after Faraday announced the theory of electromagnetic induction, Frenchman Hippolyte Pixii exhibited a hand-cranked generator. American Thomas Davenport built the first electric motor in 1834, only two years later. The invention of the modern electric generator by Anyos Jedlik of Hungary in 1861 (and by E. W. Siemens of Germany five years later) signaled the start of the electric utility industry. In 1881 Thomas Edision built the world's first central electric generating station along with a public distribution system in New York City. The construction of a 110-mile long high-tension transmission line between Lauffen and Frankfurt, Germany, in 1891 was the first step in moving electric energy to great distances.

Petroleum was produced from shallow dug wells for many years in Poland, Romania, and Russia. The first refinery was built by three Russian peasants—the Dubinin brothers—in 1823, using a crude distillation method that produced 16 buckets of kerosene from each 40 buckets of crude oil. Polish pharmacist Ignacy Lukasiewicz developed the kerosene lamp in 1852. Due to the declining whale population, by the time of the American Civil War a shortage developed of whale and sperm oil used previously for lighting. Prices rose from 43 cents a gallon in 1823 to $2.55 a gallon in 1966. This great increase in whale oil prices gave an impetus to the petroleum industry. The first drilled oil well was made by E. L. Drake in Pennsylvania in 1859. After a frantic search of petroleum fields, kerosene took the place of whale oil as an exclusive source of illumination for a period of nine years. Soon petroleum took its place in powering transportation. The improvement of the internal combustion engine by Gottlieb Daimler, and the invention of the diesel engine by Rudolf Diesel, made oil the lifeblood of transportation within a few decades after the end of the 19th century.

The growth in energy utilization in the world during the 20th century is shown in table (82) over.

The phenomenal rate of growth in the use of some of these energy carriers put great strains on production. Oil scares, much like the 1973 "energy crisis" (11), occurred in 1921, 1923, 1935, and 1947.

The increase in energy consumption by the United States seems to parallel the growth of our gross national product (GNP). For each dollar of the GNP, about 95,000 BTU's of energy were used during the past few decades. The ratio of commercial energy consumption seems to be in

Absolute values of annual production and comparison with the production of the year 1900 as a unit

	1900	1932	1950	1958	1959
Anthracite					
Million tons (metric)	701	952	1424	1819	1886
specific increase	1.0	1.4	2.0	2.6	2.7
Bituminous Coal					
million tons (metric)	72	170	382	615	620
specific increase	1.0	2.4	5.3	8.5	8.6
Petroleum					
million tons (metric)	21	181	523	927	977
specific increase	1.0	8.6	24.9	44.2	46.5
Natural gas					
1000 million m³	4	49	207	380	385
specific increase	1.0	12.3	51.8	95.0	96.3
Hydroenergy					
1000 million kWh	40	140	288	604	615
specific increase	1.0	3.5	7.2	15.1	15.4

direct proportion to the gross national product in all countries of the world, as shown in Fig. 1.2. Where the per capita energy consumption is high, so is the per capita gross national product (111). The richer a country, the more energy it consumes. The United States is by far the leading country of the world on the basis of gross national product and energy consumption (43). In fact, with only six percent of the world's population, we account for 35 percent of the world's energy consumption (31). The rate of energy utilization in the United States grows by leaps and bounds (81). Electric energy production grows with an increasing rate, doubling every ten years. Between the years of 1940 and 1969 the generation of electricity in the world grew from 505 to 4443 billion kilowatt-hours. The share of the United States was 180 billion kilowatt-hours in 1940 and 1553 billion kilowatt-hours in 1969. The total electric utility output increased over sevenfold between 1946 and 1971. With this rate of increase, Americans will be using a total of about 350,000 kilocalories per capita each day by the year 2000 (51, 54, 55).

The total amount of energy used by mankind from the dawn of history until the year 1900 has more or less increased in proportion with the population. During this century, however, the rate of growth accelerated, as shown in Fig. 1.1. This graph may be replotted on a semi-logarithmic

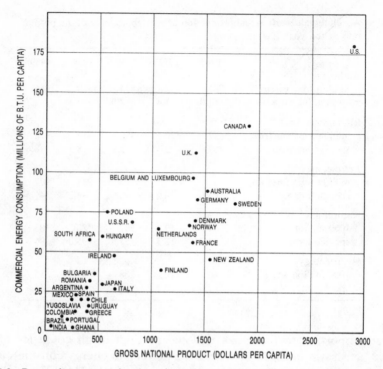

Fig. 1.2 Per capita gross national product versus energy utilization in various countries. (After Ref. 31, Copyright © 1971 by Scientific American, Inc. All rights reserved.)

scale, as in Fig. 1.3, in which case the data may fall on a straight line. Extending the line to the right would give us a forecast for the future. It shows that within a certain period of time the amount of energy used up will double. This time period is called a "doubling period" (114). The concept of the doubling period points to the frightful consequences of our fast rate of growth in the use of finite fossil fuel reserves. For example, if we have already used up 0.1% of the world's oil and the doubling period of the rate of use is ten years, the remaining oil will be gone in less than ten doubling periods—or, in other words, in less than a hundred years.

Energy use in transportation grows in the United States by a rate of 6.5% each year. The growth rate in other countries is even more staggering. The increase in crude oil consumption during a 20-year post-World War II period (between 1953 and 1972) was as follows (11):

Japan	2567%
West Germany	1597%
Italy	1079%
Benelux countries	735%
France	623%
Scandinavian countries	492%
Britain	367%
Spain	294%
Canada	270%
United States	110%

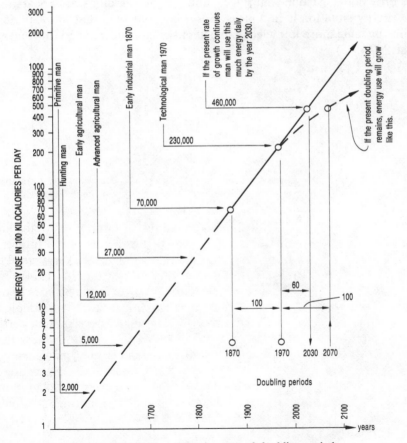

Fig. 1.3 Illustration of the concept of doubling periods.

With growth rates of such magnitude, it is no wonder that the world's supply of energy has become a central issue of our time. The energy-producing capacity of the nation is straining to keep pace with the demand (68, 119). Meanwhile, due to environmental considerations, pollution control, and public safety, the difficulty of producing enough raw energy is growing. International political difficulties create potential shortages in oil. Government regulations have an adverse effect on oil and gas production. Air pollution and anti-strip mining regulations make the utilization of coal difficult. Lack of research support retards the development of new energy production schemes; environmental concerns delayed the Alaska pipeline; etc. As a result, we are experiencing blackouts and brownouts, gasoline shortages, and fuel oil scarcities. It appears that our energy situation is in a mess. So, it is about time that the American public became more knowledgeable about what is now called the "energy crisis" (44, 45).

CHAPTER 2

What is Energy?

Energy is the capacity to do work. It is, in other words, stored work.

Energy appears in many forms. Fossil fuels—like coal, oil, or gas—contain it in the form of chemical energy. The chemical reaction we call burning releases the stored energy in the form of heat energy.

Nuclear energy is contained in the form of bonding forces within the parts of the nucleus of an atom. When heavy, unstable nuclei are broken into roughly equal parts forming new atoms, we speak of nuclear fission. Fission, like a chemical reaction, releases heat energy. Fusion is another way of releasing atomic energy; by combining (fusing) two light atoms together, a new heavier atom is formed and, in the process, energy is released in the form of heat energy.

Heat—or thermal energy as it is called—may be transformed into mechanical energy. By using the expansive property of gases or steam, gas or steam turbines transform the thermal energy into pressure or velocity. These are forms of mechanical energy.

Elevation over a base level is another form of energy—potential energy. A glass on a table has a potential energy. If it falls to the floor, this potential energy first transforms into velocity—called kinetic energy. As the glass hits the floor, its kinetic energy is used up to shatter the glass, breaking the bonding forces between its parts. When raindrops fall on a mountain, the water has a potential energy with respect to the valley, and an even greater potential energy with respect to the level of the sea. This potential energy converts into kinetic energy in the running waters. This kinetic energy inherent in the motion drives the waters back into the sea. During this process, there is friction between the water molecules

9

and the creek bed. Friction results in heat. This heat energy is lost by the water through diffusion into the atmosphere.

A compressed spring contains elastic energy. Released, it vibrates, setting the molecules of the spring into relative motion. This kinetic action again results in friction heat, which is lost into the atmosphere.

The rays of the sun contain radiant energy. Radiation is an electromagnetic vibration. The energy of the sun's radiation transforms into heat or light—both are electromagnetic radiations of different wavelengths.

Sunshine forms vegetable matter by the process called photosynthesis. By combining various chemicals into more complex compounds through the addition of energy, stored chemical energy is formed. Man and animals take this vegetable matter as food—converting it into living tissue. Therefore, both animals and man feed on stored solar energy. Fossil fuels were formed from dead plants and animals in rare favorable circumstances that prevent decay, a slow burning process.

Electric energy occurs rarely in nature; it manifests itself in the form of lightning. Man generates electric energy from either mechanical energy, by the use of turbines rotating electric generators, or from chemical energy. The latter is called "direct energy conversion" and it employs fuel cells, thermoelectric or thermionic devices.

The conversion of energy from one form to another is governed by some of the most fundamental laws of physics (20). These laws are called the first and second laws of thermodynamics. The first law, recognized by James Joule in 1840 (when he was 22 years old), states that whenever another form of energy is transformed into heat or vice versa, the energy of the form that disappears is equal to that of the other forms that appear. In other words, energy cannot be created or destroyed, although it may be lost as heat is dissipated in the atmosphere.

The second law of thermodynamics was based on the work of Sadi Carnot in 1819; it concerned ideal heat engines. This law states that it is impossible to transfer heat energy from a colder to a warmer body without doing work. To put it in another way: it is impossible to operate a heat engine that will convert heat energy from a hot body entirely into work without exhausting heat to a colder body; therefore, in any change of form there is a waste of energy.

From the second law, it follows that energy conversion is never perfect, some "waste" is inevitable. The ratio of work produced to the energy expended is called the efficiency of the conversion. Technical improvements of energy conversion systems are continually directed toward the improvement of the efficiency of the operation. Improved efficiency results in greater outputs of energy in its desired form at the original energy input. In other words, improvement of conversion efficiencies means savings in our basic energy resources.

Considering the many forms in which energy appears, it is not surprising that there are just as many and more measures for energy. In organizing the material for this book the idea occurred that it might be better to put all quantities into a standard unit. This idea was discarded in the light of the oddness of a statement like "the capacity of a supertanker is so many millions of BTU." The quantities mentioned throughout this book will be computed by their conventional unit, whatever that measure may be. To compare the data to another unit of measure, the reader may refer back to the conversion tables at the end of this chapter.

Joule is the only universally accepted measure of energy. It expresses the work done by a force of one "newton" exerted for a distance of one "meter." For example, if a box is pushed 15 feet along a horizontal floor against a frictional force of 50 pounds, the work done is 15 times 50, which equals 750 foot-pounds. The conversion from foot-pounds to joules requires that we multiply the ft-lbs by 1.356. Hence the answer to our example is 1017.0 joules. Multiples—like megajoule, meaning one million joules—are often used also.

Foot-pound expresses potential energy in a direct manner if one uses the so-called English system. It shows that the potential energy held by, say, one pound of water one foot over an arbitrary reference level is one foot-pound. The concept is, of course, identical in the metric system.

Meter-kilogram is the metric equivalent of the foot-pound. A worker who shovels 8 cubic meters of sand, weighing 1.8 metric ton per cubic meter, up a vertical distance of 1.6 meters, exerts 23,040 meter-kilograms of energy, because $8 \times 1800 \times 1.6$ equals just that. To convert meter-kilograms to joules, one has to keep in mind that one meter-kilogram equals 9.81 joules. (One must know here that kilogram in common practice is used as a force acting at the surface of the Earth. There the gravitational acceleration of 9.81 meters per second^2 acts on a mass of one kilogram. We have to be conscious of the important difference between force and mass. To distinguish between them, the term "newton" is used for force and the term "kilogram" is used for mass. By the law of Newton, force equals mass times acceleration, the latter being that of the gravitational pull in the case of weight. Of course, weight disappears in outer space where there is no gravity, while mass is constant everywhere. In the English system the similar distinction is made by the little known "poundal" term, which equals 0.1382 newton.)

BTU ("British Thermal Unit") is the English way of expressing heat energy. It is the amount of heat required to raise the temperature of one pound of water by one degree Fahrenheit. One BTU equals 107.5 kilogram-meters, or 1055 joules.

Calorie is the metric measure of heat energy. It is the energy required to heat one gram of water by one degree Celsius (centigrade). Calorie is a small quantity because it refers to the small mass of one gram. Instead of this, kilogram-calorie is used. It refers to heat energy required to warm one kilogram of water by one degree Celsius. In short it is called kilocalorie. One BTU is equal to 252 calories; conversely, one calorie equals 0.00397 BTU. One kilocalorie is therefore 3.97 BTUs. To convert to joules, one must multiply kilocalories by 4186. Fossil fuels are usually converted into heat energy. Their measure is always based on their physical size rather than their heat energy content.

Coal is measured by the ton. Anthracite coal contains 23,400,000 BTUs per ton. Hence, anthracite coal holds 24,700,000,000 joules, a large number indeed. Engineers and scientists do not like such large numbers lest they lose some of the zeroes. Usually large numbers are written in the "exponential form." We will find this throughout this book. The energy in anthracite will be expressed in the form of 2.47×10^{10}, or 247×10^{12}, and the like. This simply means that after writing the first quantity it must be multiplied by ten as many times as the exponent over the ten indicates. In short, one has to move the decimal point to the right as many spaces as the superscript indicates. We save the great many zeroes by this notation. Also, the exponent is by far the most important part of the number. Most people, including scientists, have difficulty imagining these large numbers. Only a few people can think in terms of over 100,000, or 10^5. Some examples may aid our imaginations.

Imagine the size of one cubic foot. The area of the state of California is about 4.38×10^{12} square feet. Covering all of California with 10 feet of water needs about 4.38×10^{13} cubic feet of water. Flooding California with water to the height of the Empire State Building (1250 feet) would require about 5.47×10^{15} cubic feet of water. Covering California with enough water to reach the top of Mount McKinley (which is still in Alaska at this time), 20,320 ft depth would be needed. The amount of water in cubic feet would equal about 8.9×10^{16} cubic feet or 3.5×10^{19} in cubic inches. These numbers are comparable to those used in this text. One has to recognize that the large quantities are caused by the small basic units used. If, for example, BTU would represent the heat to warm up one cubic mile of water by 100°F, our need for these large numbers would disappear.

Oil is measured by the barrel in the United States. One barrel of oil contains 42 gallons*. The heat energy content of oil is 5,800,000 BTUs per barrel. This equals 6.1×10^9 joules per barrel or 0.145×10^9 joules per gallon. In the metric system, oil is measured by the metric ton. One metric ton of oil is 7.22 barrels. Hence, one metric ton (1000 kg) of oil contains 44×10^9 joules of energy.

Natural gas is measured by the cubic foot. As gas is compressible, measuring it by volume is not a particularly fortunate choice, but this is the way it is done. In fact, to eliminate the variations of mass in the volume by changes of temperature and pressure, a standard is defined**. The energy content of a standard cubic foot of dry natural gas is 1035 BTUs per cubic foot. This is about 1.09×10^6 joules per cubic foot. The metric system uses cubic meter as a unit. A cubic foot is 0.0283 cubic meter; therefore, 35.31 cubic feet fills one cubic meter. The energy content of one standard cubic meter of gas is, then, 38.6×10^6 joules.

Solar energy is measured by the langley. One langley is the radiant energy of the sun in calories falling on an area of one square centimeter in one minute. The mean value of this radiation in the continental United States is about 500 to 700 calories per square centimeter daily on each typical summer day. To convert langleys to joules, one must multiply langleys by 4.186, getting the resultant in the form of joules per cm^2 per minute.

Geothermal energy deals with heat capacity, measured in BTUs per cubic foot. One BTU per cubic foot equals 8.92 calories per cubic meter or 1055 joules per cubic foot of hot rock.

Power is derived from energy. It is different from work or energy because it also involves the time during which the work is done. To lift a 500-pound steel girder to the top of a 100-foot high building, the same amount of work (50,000 foot-pounds) is done, regardless of the time it takes to perform the task. This is power expressed by work divided by the time during which it is performed. The performance of the above task in two minutes requires a power of 25,000 foot-pounds per minute, while doing it in ten minutes requires only one-fifth of that power—5000 foot-pounds per minute. Common English terms for power are the foot-pounds per second and the horsepower. One horsepower (HP) equals 550 foot-pounds per second or 33,000 foot-pounds per minute. In the metric system the horsepower is only slightly different. One English

---

*United States gallons are used throughout this book. There are 35 imperial gallons in a barrel.

**E.g., 60°F temperature and 6 ounces per square inch over atmospheric pressure used by gas utilities in the United States.

Energy

	BTU	erg	ft-lb	hp-hr	joule	cal	kw-hr	EV	MEV
1 BTU	1	1.055×10^{10}	777.9	3.929×10^{-4}	1055	252.0	2.930×10^{-4}	6.585×10^{21}	6.585×10^{15}
1 erg	9.481×10^{-11}	1	7.376×10^{-8}	3.725×10^{-14}	10^{-7}	2.389×10^{-8}	2.778×10^{-14}	6.242×10^{11}	6.242×10^{5}
1 ft-lb	1.285×10^{-3}	1.356×10^{7}	1	5.051×10^{-7}	1.356	0.3239	3.766×10^{-7}	8.464×10^{18}	8.864×10^{12}
1 hp-hr	2545	2.685×10^{13}	1.980×10^{6}	1	2.685×10^{6}	6.414×10^{5}	0.7457	1.676×10^{25}	1.676×10^{19}
1 joule	9.481×10^{-4}	10^{7}	0.7376	3.725×10^{-7}	1	0.2389	2.778×10^{-7}	6.242×10^{18}	6.242×10^{12}
1 cal	3.968×10^{-3}	4.186×10^{7}	3.087	1.559×10^{-6}	4.186	1	1.163×10^{-6}	2.613×10^{19}	2.613×10^{13}
1 kw-hr	3413	3.6×10^{13}	2.655×10^{6}	1.341	3.6×10^{6}	8.601×10^{5}	1	2.247×10^{25}	2.270×10^{19}
1 EV	1.519×10^{-22}	1.602×10^{12}	1.182×10^{-19}	5.967×10^{-26}	1.602×10^{-19}	3.827×10^{-20}	4.450×10^{-26}	1	10^{-6}
1 MEV	1.519×10^{-16}	1.602×10^{-6}	1.182×10^{-13}	5.967×10^{-20}	1.602×10^{-13}	3.827×10^{-14}	4.450×10^{-20}	10^{6}	1

Power

	BTU/hr	ft-lb/sec	hp	cal/sec	kw	watt
1 BTU/hr	1	0.2161	3.929×10^{-4}	7.000×10^{-2}	2.930×10^{-4}	0.2930
1 ft-lb/sec	4.628	1	1.818×10^{-3}	0.3239	1.356×10^{-3}	1.356
1 hp	2545	550	1	178.2	0.7457	745.7
1 cal/sec	14.29	3.087	5.613×10^{-3}	1	4.186×10^{-3}	4.186
1 kilowatt	3413	737.6	1.341	238.9	1	1000
1 watt	3.413	0.7376	1.341×10^{-3}	0.2389	0.001	1

horsepower equals 76.05 meter-kilograms per second. Needless to say, horsepower has little to do with horses, except that in 1782 Watt found that a "brewery horse" could produce 32,400 ft-lbs per minute. Another way of writing power is as the product of force and velocity. The force exerted by the propellers of a ship to overcome resistances of the water, multiplied by the speed of the ship, results in the power of the ship's engines (less losses due to the lack of 100% efficiency in the whole operation).

Electric power is commonly measured in watts. One watt of power is one joule of energy per second. Since this is a very small quantity, electrical engineers use 1000 watts, called kilowatts. Electric utilities use even larger units. A megawatt is one million watts. A gigawatt is even larger—1000 million watts, or 10^9 watts. So a gigawatt is 10^9 joules of energy per second. One kilowatt is 1.34 horsepower.

Electric energy is expressed in kilowatt-hours or megawatt-hours. The electric bill of homeowners is based on the amount of electric power used over a period of time. For example, using a 100-watt light bulb for 10 hours uses up 100×10 equals 1000 watt-hours, which is a kilowatt-hour of electric energy. To convert kilowatt-hours into BTUs, the former must be multiplied by 3415. One kilowatt-hour of electric energy equals 2.655×10^6 foot-pounds or 860.5 kilocalories or 3.671×10^5 meter-kilograms.

Nuclear energy is used to make heat in the place of fossil fuel-fired boilers in the generation of electric energy. One gram of uranium-238 contains enough energy to make 8.1×10^{10} joules of heat. The term "electron volt" is also used in the field of atoms. One million electron volts, MEV, equals 1.6×10^{-13} joule—a very small number. One billion electron volts, BEV, equals 1.6×10^{-10} joule.

The tables opposite show some of the most common conversion factors related to energy and power.

CHAPTER 3

Ways We Use Energy

To give an example of the amount of energy America uses each year, one should only imagine that it equals the energy contained in 2700 gallons of gasoline for each individual man, woman, and child in the United States. This amount of energy is used up in the form of gasoline that runs the tractor on the farm, and in the trucks that carry those farm products to the market. Some of it is used to make the steel, to make the tractor and the truck, to make the machines that build the roads. This is the energy that is used by the worker in the factory while earning his money to pay for his food, for his home, to keep up with the electric bills, his heating bills, and to keep his automobile running. In fact the average American industrial worker uses over 46,000 kilowatt-hours of electricity on his job every year. The work of this much electricity equals the manpower of about 700 men. In contrast to his counterpart of 150 years ago, who essentially worked without any energy help except an occasional horse, an American worker has 700 helpers each day on his job (50).

Although much of the energy consumed in the United States is actually used up in industry (producing items of industrial and personal use), about a third of the energy requirement is used up personally by Americans. Heating, air-conditioning, light, cooking, drying, and water heating all require energy. The myriads of convenience items, electric toothbrushes, electric knives, typewriters, razors, etc. not only cost energy during their manufacture but use energy as well. In 1963 only 19% of American families had air-conditioners, ten years later this figure rose to 47%. During the same decade, families with electric blankets have doubled. Color TVs, heavy users of electric current, were found in 5% of American

17

homes; today this number is 61%. Not only do we use more electricity but there are more of us using it. In 1962 there were 53.7 million wired homes. In 1973 there were 67.3 million.

The United States consumed 64.6×10^{15} BTU energy in the year 1970 (31). The sources of this energy were as follows:

coal		13.5×10^{15} BTU	20.9%
oil	domestic	17.1×10^{15} BTU	26.5%
	imported	6.8×10^{15} BTU	10.5%
gas	domestic	23.4×10^{15} BTU	36.2%
	imported	0.9×10^{15} BTU	1.4%
water power		2.7×10^{15} BTU	4.2%
nuclear power		0.2×10^{15} BTU	0.3%
TOTAL		64.6×10^{15} BTU	100%

These data are also shown in Fig. 3.1.

Of this total energy input, 17.0×10^{15} BTU was used up in the production of electricity (31). Of this total amount, the share of the various energy resources was as follows:

coal	7.8×10^{15} BTU	46.0%
oil	2.3×10^{15} BTU	13.7%
gas	4.0×10^{15} BTU	23.6%
water power	2.7×10^{15} BTU	15.9%
nuclear power	0.2×10^{15} BTU	0.8%
TOTAL	17.0×10^{15} BTU	100%

Of this total energy producing electricity in that year 11.8×10^{15} BTU was lost in the form of wasted heat during the processes of generation and transmission of electricity. This amounts to 69% of the production (31). Only 31%, amounting to 5.2×10^{15} BTU, was actually delivered to the users. These users were industry—with a consumption of 2.3×10^{15} BTU—and commercial and residential users—using up 2.9×10^{15} BTU.

The rest of the total energy was used directly without first converting it to electricity. This energy amounted to 47.6×10^{15} BTU, 73.7% of the total. The users of this energy were

residential and commercial	12.9×10^{15} BTU	27.1%
transportation	16.3×10^{15} BTU	34.2%
industrial users	18.4×10^{15} BTU	38.7%
TOTAL	47.6×10^{15} BTU	100%

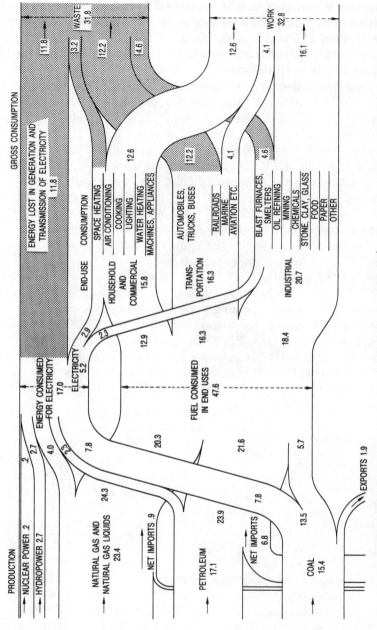

Fig. 3.1 Various sources and uses of energy in the United States in 1970, expressed in 10^{15} BTU's. (After Ref. 31, Copyright © 1971 by Scientific American, Inc. All rights reserved.)

19

In the field of transportation 16.3×10^{15} BTU was used up in the form of gasoline and diesel fuel-run automobiles, trucks, boats, and airplanes. Of this total, about 25.2% did useful work—4.1×10^{15} BTU. The remainder (12.2×10^{15} BTU) ended up as waste heat, a staggering 74.8% of the total energy used in transportation.

Industrial users of energy bought 20.7×10^{15} BTU worth of fuel, plus 2.3×10^{15} BTU electricity. Of this total energy, 16.1×10^{15} BTU did useful work while 4.6×10^{15} BTU was wasted—20% of the total energy purchased by industry.

All in all, in the year 1970 (used for this example), the United States converted 32.8×10^{15} BTU energy into useful work while 31.8×10^{15} was wasted in the process. In these calculations, we neglected significant amounts of energy resources that were left behind during extraction. These amounted to perhaps 7×10^{15} BTU of oil that cannot be brought to the surface from existing oil fields with present technology, and perhaps an equal amount of coal left in the ground in smaller seams or on waste piles, etc. With these uncertain amounts included, we can estimate that in 1970 we wasted perhaps 150% of the energy resources actually used. Therefore, we can conclude that our energy utilization efficiency ranges well below 50% and may be as low as 30% (46).

Efficiency is the energy (or power) output divided by the energy input, expressed as a percentage. For example, a thermal power station that furnishes 100 megawatts of electricity by using 300 megawatts of thermal energy in the form of coal is operating at an efficiency of 33% (33).

The present efficiency of various energy converting devices (114) is shown below:

incandescent lamp	5%
steam locomotive	8%
solar cell	10%
thermoelectric converter	13%
thermionic converter	15%
rotary engine (Wankel)	18%
fluorescent lamp	20%
internal combustion engine	25%
solid state laser	30%
steam power plant with generator	32.5%
high intensity lamp	33%
industrial gas turbine	34%
aircraft gas turbine	36%

nuclear reactor	39%
steam turbine	46%
liquid fuel rocket	47%
steam power plant with MHD generator	50%
fuel cell (hydrogen-oxygen)	60%
electric storage battery	72%
home gas furnace	85%
dry cell battery	90%
hydraulic turbine	92%
electric motor	93%
electric generator	98%

Most of these efficiency factors are averages within a range of $\pm 10\%$.

Much of the loss is unavoidable. With the state of today's technology we are doing the best we can. There were tremendous improvements in the efficiency of energy utilization during the past decades. The overall improvement of the efficiency of consumption improved by about a factor of four since the turn of the century. One of the best examples of this is the efficiency of lighting. Light output is commonly measured in lumens. Ordinary incandescent lamps at the turn of the century operated at a 1% light efficiency. A 100-watt light bulb in 1900 turned out about 1 watt of visible light and 99 watts of heat. Today this has improved to over 5 watts of light output or 11 to 22 lumens per watt, depending on the "color"—the light wave frequencies of the produced light. Luminous or lighting efficiency of today's lamps are as follows:

incandescent	11 to 22 lumens/watt
fluorescent	73 to 80 lumens/watt
mercury vapor	45 to 55 lumens/watt
metal-halide	80 to 100 lumens/watt
tungsten-halogen	19 to 22 lumens/watt
high pressure sodium	102 to 130 lumens/watt

If one were satisfied with a light of a single wavelength of peak sensitivity by the human eye, theoretically a maximum of 680 lumens per watt would be attainable. A broad mixture of various wavelengths, on the other hand, is required to produce white light; this could be produced at a maximum of about 220 lumens per watt, theoretically. It is assumed that about 2% of the total energy of this country is used for lighting. An improvement in lamp efficiency could, therefore, result in a considerable energy saving. This could come about by investing large amounts of research monies. An alternative, of course, is to go around turning off lights (48).

Much of the electricity used by an average American is not used for lighting. In fact, such use is only about 0.9% of the energy used in an individual household. In 1970 the average American used 31.5% of his share of energy for space heating and 6.8% for water heating; 41.5% of the energy used by this average American was spent on running the family car, 2.6% was used for cooking. The rest went for air-conditioning (0.9%—up 200% during one decade) and other minor uses.

Many home appliances use electricity. The table below shows the electric power requirement of various common household applicances and their average yearly power consumption.

	Average Wattage	Kilowatt-hours Consumed Annually
Comfort		
air-conditioner (room)	1,566	1,389
electric blanket	177	147
dehumidifier	257	377
fan (rollaway)	171	138
heater (portable)	1,322	176
heating pad	65	10
humidifier	177	163
lighting	—	1,800
Health		
hair dryer	381	14
shaver	14	1.8
sun lamp	279	
toothbrush	7	0.5
Entertainment		
radio	71	86
radio-record player	109	109
television (black and white)	237	362
television (color)	332	502
Housewares		
clock	2	17
floor polisher	305	15
sewing machine	75	11
vacuum cleaner	630	46

	Average Wattage	Kilowatt-hours Consumed Annually
Food		
blender	386	15
broiler	1,436	100
carving knife	92	8
coffee maker	894	106
deep fryer	1,448	83
dishwasher	1,201	363
freezer (frostless, 15 cu ft)	440	1,761
frying pan	1,196	186
mixer	127	13
oven, microwave	1,500	300
oven, self-cleaning	4,800	1,146
range	8,200	1,175
refrigerator (frostless, 12 cu ft)	321	1,217
toaster	1,146	39
waffle iron	1,116	22
waste disposer	445	30
Laundry		
clothes dryer	4,856	993
iron (hand)	1,008	144
washing machine	512	103
water heater (standard)	2,475	4,219

As shown from the data above, there are only a few major users of electric energy in the home. These are the heating and cooling systems, oven and refrigerator-freezer, and water heater. These add up to over two-thirds of the electric energy in the "all-electric home" not counting the biggest single users, the electric heating system and lighting.

Much of the energy used for space heating is wasted because of the poor insulation of most buildings. Perhaps a 30% saving could be realized in heating costs by the improvement of the insulation of the average American home. As about 80% of the heat loss occurs through the ceiling, a standard national building code requirement of at least six inches of ceiling insulation could save enormous amounts of energy that is used today to melt snow off the roofs (29). The technology is available to save much of the energy lost in home heating. New insulating materials exist that could keep a cup of coffee warm for 2000 years. Recent studies indicate that by better insulation, smaller window areas, and changing the floor shape of buildings to a square, a saving of about 35% could be realized in heating costs. Another 16% saving could be gained by improving the design and the control system of the heating, ventilating, and air-conditioning systems of buildings.

Heating with electricity is a wasteful process. Electric current is high-grade energy. We burn fossil fuel at several thousand degrees to create electricity, only to bring back heat for low-grade use, heating air to 70°F and water to 140°F. The average efficiency of electric energy production requires about three BTUs of input for each BTU of output, an expensive energy from the standpoint of heating homes. In some applications heat pumps may be considered as exceptions to this statement, but unfortunately few heat pumps are used in homes in spite of the fact that they have been commercially available for many years. Thanks to the 1973 oil scare, a large number of the homes in America may be heated in the future by solar heaters that are now under development.

Not all energy savings are made at the point of use. The example of water heating in the home may be the most illustrative one (31). Electric water heaters are claimed to be 100% efficient, but to warm 50 gallons of water to a temperature of 212°F (from 32°F) they require 75,000 BTUs of heat. A gas water heater wastes about 44,000 BTUs through the chimney, but the production of 75,000 BTUs of electricity requires the waste of 150,000 BTUs of heat at the electric-generating plants. Neglecting transmission costs in both cases, we find that the electric water heater needs 234,000 BTUs of raw energy while the gas water heater needs only 121,000 BTUs—almost 50% less as shown in Fig. 3.2. The same concept applied to space heating results in a technical efficiency of 75% for direct fuel use (gas heat) versus 32% efficiency for electric heat in the American TV dream of an "all-electric home" (30).

Commercial buildings, notoriously overheated in the winter and over-cooled in the summer, are perhaps the most wasteful structures. In addition to the tremendous heat losses incurred through the glass walls, the cooling and heating systems are so designed that about three or four times as much air as necessary for public health purposes is sucked into the building. This air is then heated in the winter and cooled in the summer. On a warm winter day, when an open window or two would provide comfort, the automatic controls of these heating systems turn on the air-conditioning.

A new trend in improving the efficiency of space comfort systems is what is referred to as a total energy plant. This is a miniature electric-generating plant incorporated into the building design. There are several thousand such systems already operating in the United States. The fuel is taken directly to

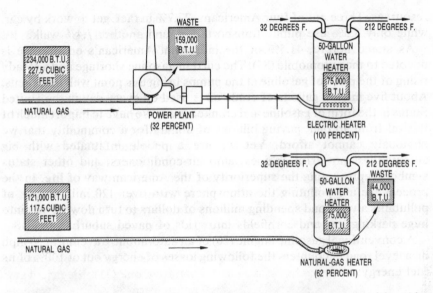

the building, generating electricity at the site. The heat loss of the electric generating process is then used in the heating or air-conditioning of the building. The overall improvement of efficiency is about 30%. This process was incorporated into the original design of the World Trade Center in New York City but (by offering a package of promotional rates) the local electric utility successfully influenced the owners to adopt conventional systems. As a result the World Trade Center has the dubious reputation of being one of the biggest energy-wasters in the country, using electricity at a rate equal to that of Stamford, Conn., or Schenectady, N.Y.

A major step toward recycling energy wasted in the electric generating process was made in Budapest, Hungary where a large portion of the city has been heated centrally from an electric generating plant since the early sixties. In Germany a nationwide central heating system utilizing the excess heat of nuclear reactors and conventional generating plants is planned to save 80 million metric tons of coal each year by the end of this century. This amount is about equal to the total 1970 coal production of Germany.

Transportation is a major use of energy. Of all public and private transportation modes (cars, trucks, railways, buses, airplanes, boats, and farm vehicles), the private automobile uses about half of the available resources. Three out of four Americans, 77.7% in fact, get to work by car, while only 8.5% use public transportation, and another 7.4% walk.

As stated before, 41.5% of the individual American's energy use is devoted to the automobile (85). The current gasoline shortage and a steady rising of the price of gasoline at the pumps bring this point well into focus. About five million gallons of crude oil is used up every day in the United States in the form of gasoline and crankcase oil. We have to import much of this oil from abroad, paying billions of dollars for a commodity that we obviously cannot afford. Yet we are a people infatuated with big automobiles, powerful engines, auto air-conditioners, and other status symbols representing the superiority of the American way of life. In the process we are polluting the atmosphere with over 120 million tons of pollutants a year, and spending millions of dollars to turn downtowns into huge parking lots and cornfields into grids of paved suburban roads.

A conventional gasoline engine (20) of an automobile traveling at 30 mph on a level road encounters the following losses of energy out of 100% of its fuel energy:

Energy losses incurred in converting to mechanical energy:

cooling water	−35.8%
exhaust gases	−35.6%
exhaust pipe	−1.0%
muffler	−1.2%
engine friction	−5.6%

Therefore, the engine's remaining energy is 20.8%.
Energy losses due to the motion of the car are:

rear tires	−3.7%
front tires	−1.1%
front wheels	−0.6%
air resistance (average)	−7.1%

Hence, the car's excess power to accelerate, climb, etc. is only 5.4%.

It may be noticed that most of the energy is wasted in the cooling system and through the exhaust gases, less than 5% is used to overcome road friction. Hence, the gasoline engine is an eminently inefficient energy converter. In spite of the steady improvements in all fields of technology, the gasoline consumption of the average American car decreased from over 13 miles per gallon before 1950 to under eight miles per gallon in 1973.

Some of this loss in efficiency is understandable. Emission control devices required for 1973-model automobiles increase the gasoline consumption by as much as 7%. The typical automatic transmission increases the fuel consumption by 6%. An increase of 500 pounds in car weight reduces the mileage by about 14%. A 5000 lb automobile uses twice as much gasoline as a 2500 lb one. Air-conditioners reduce the gasoline mileage by 9% on the average, but on a hot summer day, in city traffic, they increase the gasoline consumption by as much as 20%. On the other hand, a steel-belted radial tire can cut consumption by 10%. Keeping tires properly inflated, engines tuned, buying no more than 91 octane fuel (all cars are designed to run on "regular" since 1971), and avoiding excessive acceleration and speed can cut gasoline consumption significantly. Even greater cuts in gas consumption can be attained by not using the car unless it is absolutely necessary. Car pools, use of public transit, and the avoidance of "Sunday driving" would tremendously improve our gasoline supply situation. So would a progressive taxation of automobile by weight (an old practice in Europe), a limitation of automobile size and weight by legislative means, and a curtailment of gasoline sales either by increased prices or by strictly enforced rationing. These steps would certainly be unpopular for most citizens, but so is the devaluation of the dollar, the unfavorable balance of trade, the sinking of American influence abroad, the inflation, the food crisis, and other problems that beset the country (48).

While we are fighting to keep our internal combustion engines running, engineers are trying to devise ways to turn to a more efficient and less polluting mover. Alternatives to the standard reciprocating engine are the rotary (Wankel) engine, now well-known in the Japanese Mazda, the stratified charge engine of the Honda, (also Japanese), the gas turbine engine, and the diesel engine used by Mercedes, Peugeot, and truck manufacturing companies (6). While most of these show definite advantages in at least some respects over the common piston engine, their overall evaluation still leaves a lot of advantages on the side of the conventional engine. Clearly, there must be a great deal of further research before the conventional engines are finally off the road.

Alternative power generation for automobiles is also under study. The steam-operated Rankine engine, the Stirling engine that uses helium and hydrogen as a working fluid, and the electric car all show at least some potential future application. The electric car deserves a great deal of attention. The development of more efficient rechargeable batteries may allow a new efficient "second car" to come on the roads—one that would allow the driver to travel to work and back home on one charge and could then be recharged overnight. Perhaps in the distant future, when electrical energy transmission by microwaves to cars on the road will be a feasible reality, we will have some other alternatives (6).

For the foreseeable future, we are unlikely to see a change in engine configuration. General Motors alone has 12 final engine assembly lines. Backing these are a total of 114 major production lines, producing subassemblies for engines. To convert even one of these would take several years and many millions of dollars. Why should an automobile manufacturing company spend this kind of money as long as it is not required to do so, as long as the oil companies are more than happy to furnish the gasoline (if available somewhere in the world), and as long as the American consumer wants a new model car every year, loaded with all the extras (91)?

Fundamentally, the possibility of unlimited individual transportation afforded by the automobile was instrumental in shaping practically all our present day environment from the decay and virtual abandonment of the inner city to the rapid creation of suburban areas. Changing this living pattern could be attained by the changing mode of transportation. But, as it has taken over a half century to establish the type of environment we are accustomed to today, it is prudent to say that changing it again will take at least as much time. The automobile, therefore, will be with us for a long time. It may become smaller, its use may be limited in the inner cities after other modes of public transportation techniques become prevalent, but the fundamental concept will change but little. From an energy standpoint, we can only hope that by improved technology the operation of automobiles will improve in efficiency from the present 25% to somewhat more. This will take much research.

In the future the high cost of transportation may force the nation to consider this aspect in the planning of our environment. Cluster housing, a return to the cities by restrictive suburban zoning, and governmental encouragement of urban redevelopment in order to move the population back into the cities would be needed. Making the American city a less undesirable place to live would go a long way toward this aim. Spending one hour each working day on an endless strip of concrete, surrounded by a cloud of carbon monoxide and a herd of charging dinosaurs should not necessarily be the required lifestyle of millions of Americans. We have spent tens of billions on our magnificent interstate highway system, hardly any was spent on biketrails in the urban and suburban areas. Perhaps one or two percent of our highway expenditures would have been enough to entice five or ten percent of the commuting public to take to bicycles.

For longer distances, railroads are much safer and more efficient than highways. A six-lane highway can move 9000 people per hour with an average car occupancy of 1.2 person per trip. A single railroad track can transport 60,000 people per hour. Traveling in an electric-powered train is 23 times safer than by car, two and a half times safer than by airplane. It is also less polluting, cheaper, and more comfortable. We may look either to Japan or Europe to find ample proofs for these facts. The energy savings implicit in train transportation are enormous. To move one ton of freight by an airplane requires 42,000 BTUs. A truck requires 2800 BTUs. But a train needs only 670 BTUs. The savings in land are also considerable. In Los Angeles over 60% of the land area is devoted to automobiles. Rapid transit would need much less, freeing land for more useful purposes.

Even more revealing is the relative cost of the different modes of personal transportation. In Germany, where the various public transportation systems are well developed, studies in 1973 indicated that the relative energy uses (relative to moving one person 100 kilometers by automobile) are as follows:

automobile	1.00
airplane	1.53
train	0.68
streetcar	0.50
bus	0.26

These numbers indicate that going by plane requires 50% more energy, while going by bus requires only one quarter of the energy that it takes to get somewhere by car. Surveys made by the Motor Vehicle Manufacturers Association indicate that Americans traveled over one trillion miles by car in 1973. Seventeen percent of this mileage involved trips of 100 miles or more one way, yet it represented less than one percent of the individual trips made by each car. This shows the tremendous savings of gasoline if, for example, car trains would be made popular in the United States or if intercity travel by car would be somehow curtailed.

The other major factors in energy conversion are fundamentally technological (114). A major user of energy is industry. This use is relatively efficient, since energy used is a direct cost of production and this cost factor is fundamental in competitive pricing.

The thermal efficiency of work improved in industrial processes from a mere 2% in 1850 to over 20% in 1970. Further improvements will not be as spectacular. Modern industrial processes are energy intensive, therefore, overall utilization will rise to a great degree. With the introduction of more efficient industrial equipment and technology, energy needs are somewhat reduced. But as the growth of production increases, these savings quickly disappear. To produce a ton of aluminum, 60 million BTUs are needed. Steel requires only 26 million BTUs; glass still less—a mere 17.5 million BTUs per ton. Hence, it would be better to use glass for containers, although the present trend is to use throw-away aluminum cans rather than returnable glass bottles.

Recycling useful waste could cut the expenses of manufacturing. To manufacture low-grade paper from recycled paper instead of virgin pulp requires 70% less energy. Making steel from scrap metal instead of ore requires 74% less energy.

Urban waste and garbage is presently produced at a yearly rate of about one ton for each person in the United States. Much of this contains recycleable products, most could be burned in incinerators, resulting in only a small fraction that must be disposed of in sanitary landfills. The heat can be used to produce steam used to warm buildings in the central city—this is already practiced in Europe and in a few American cities.

Producing methane gas from our municipal, industrial, and agricultural wastes is one way to substitute our dwindling gas supplies. Based on our present rate of waste production, we could generate methane gas with existing technology to complete favorably with natural gas. Controlled digestion without the presence of oxygen is commonly used in sewage plants to reduce organic wastes. Today much of the methane produced there is burned.

Solid waste production in the United States is estimated to be 2.6×10^{12} pounds each year. This would produce some 26×10^{12} cubic feet of methane per year, a quantity comparable to our present rate of natural gas use. This, of course, could hardly be produced realistically, but it gives an indication of the scope of these resources.

Municipal solid wastes are beginning to be used to produce steam by burning in specially designed facilities. After shredding, the waste is separated and some constituents—such as glass, aluminum, copper, and iron—are recycled.

The shredded burnable waste is burned in what is called "semi-suspension firing," which reduces the waste into a small amount of ash that is easily disposed of. The resulting steam and hot water affords savings of other fuels such as coal and gas. A midsize (half-million population) American city can generate over 300 million pounds of steam year round from its solid refuse. This and similar types of resources could well take part of the load off our rapidly burned natural gas resources.

Incinerating collectible trash could produce about 3.5% of the energy requirement of the United States. Paper waste has almost as much heating value as coal, while the various plastics found in trash contain twice as much. From the combustion of its trash, a city could produce perhaps as much as 10% of its electric needs.

There remains, as an area worthy of improvement, the electric utility industry. Here the efficiency has improved from 5% in 1900 to about 33% in 1970. The waste heat generated in electric power stations would be more than enough to heat all homes in the United States. The better nuclear reactors that will carry much of the load by 1980 operate at about 40% thermal efficiency. Theoretical limitations will make it impossible to improve this efficiency further. Transmission losses will improve in the future if the development of low-temperature "cryogenic" underground transmission lines are successful, but they will add little to the overall efficiency of the production of electricity. Perhaps a vigorous effort to develop independent electric energy producers, wind generators, fuel cells, and the like will make it possible to supply electricity to homes outside the high-density urban areas without the need of the present web of transmission lines. These and other similar developments, such as solar heating, will allow us to develop a new lifestyle by the end of the century—a lifestyle based on convenience, comfort, and freedom.

CHAPTER 4

Coal

Coal is one of the most important raw materials of modern industry. It is of organic origin, formed by the decaying vegetation at the bottom of swamps. As the trees, plants, and algae die in the semitropical swamps, they accumulate on the bottom—forming a layer of peat. When this peat is covered by the accumulating clay and sand, the weight of these squeezes out the water and compacts the organic deposit (17).

Overlying pressure due to successive soil layers deposited by water on the top of the bed of peat squeezes out even more of the water and this process forms a layered pulpy mass intermingled with fragments of wood. This is called lignite. Further increase of density results in layered black, shiny so-called bituminous coal. This coal is very abundant in the United States and elsewhere and is used for purposes of heating and iron manufacturing. Additional pressure, usually caused by horizontal pushing that is present in the mountain-forming process, causes the bituminous coal to transform into anthracite—a high-ranking coal characterized by its jet-black color, high-fixed carbon content, and its dull-to-brilliant or submetallic luster. The Appalachian mountain region is the chief supplier of anthracite in the United States (42).

It is estimated that 20 feet of dead plant matter compacts to three feet in the formation of peat. Twenty feet of dead plant matter will take about 3000 years to accumulate under the conditions prevailing today in a southern Florida swamp. If the geological conditions would change, causing a slight lowering of the ground, that would result in the deposition of clay and sand from the water to cover this peat layer. Then this three-foot layer of peat would compact into a layer of one foot of bituminous coal. Therefore, a

33

10-foot coal bed would require peat accumulation for 30,000 years and a 20-foot coal bed would require about 60,000 years. Plants and other vegetation are formed from water and from carbon dioxide, which is present in the air. The forming process is called photosynthesis, which requires the energy of the sun. On the average the sun's rays deliver 1.94 calories per minute of energy on each square centimeter of the earth surface*. To form one molecule of coal, 4200 calories of solar energy are required together with 6 molecules of CO_2 (carbon dioxide) and 5 molecules of water. When burning, the reaction is reversed, and the 4200 calories of energy are released. To form coal in large quantities, sufficient carbon dioxide must be available in the atmosphere. This happened in certain geological ages when the crust of the Earth was in motion and mountains were formed (14). During those times, there was violent volcanic activity releasing CO_2 into the atmosphere, as well as slow sinking of coastal marshy areas creating swamps—both conditions necessary to form peat. There are coal fields in which many layers of coal alternating with clay seams are present, indicating alternating sinking and rising of the land. In the coal regions on the border of France and Belgium there are sometimes 400 such coal seams on top of each other, indicating as many successive coal-forming periods.

Most coals are formed by the process of slow sinking of swamps while some are formed by water transporting organic substances into ponds or deltas. The coal fields of Commentry and Decazeville in France (17) are examples of the latter type.

Coal-forming conditions were particularly good during the carboniferous age of geologic history, about 350 million years ago. The earliest known deposits of coal are as much as 2600 million years old.

Energy in coal is present in the form of carbon and hydrogen. Depending on the type of coal, different amounts of carbon are present. Anthracite may have 96% fixed carbon, while lignite has as little as 38%. The calorific (heating) value of anthracite coal is 25,400,000 BTU/ton, while bituminous coal has 26,200,000 BTU/ton and lignite has 14 to 20 million BTU/ton.

Worldwide distribution of coal shows a large concentration in the Northern hemisphere, as shown over.

Mineable coal is defined as approximately 50% of the total, hence the total amount of coal present on the earth is about 15,300,000,000,000 metric tons. This is in layers not smaller than 14 inches (36 cm) located at depths no deeper than 1.2 km (4000 feet). Figure 4.1 shows the location of US coal fields.

*More on this in Chapter 8.

Mineable Coal Resources of the World (62)	
United States	$1,486 \times 10^9$ metric tons
North America outside US	601×10^9 metric tons
Western Europe	377×10^9 metric tons
USSR (including European part)	$4,310 \times 10^9$ metric tons
Asia (excluding USSR)	681×10^9 metric tons
South and Central America	14×10^9 metric tons
Oceania (including Australia)	59×10^9 metric tons
Africa	109×10^9 metric tons
TOTAL WORLD RESOURCES	$7,637 \times 10^9$ metric tons

The rate of coal production is now about 3 billion metric tons per year. A train holding this much coal would encircle the Earth about 15 times. It was established that if our yearly production does not reach phenomenal rates, we could expect our coal reserves to last hundreds of years. With our increasing technology, it is expected that the rate of coal production will rise considerably in the future. Estimates of yearly coal use show that the peak production will be reached between the years of 2100 and 2150. After

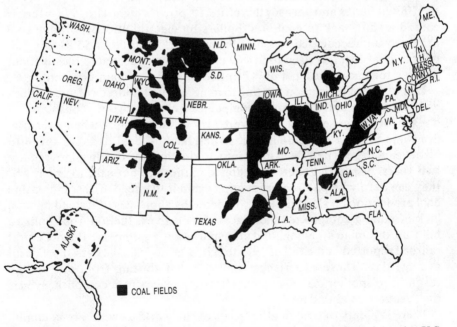

Fig. 4.1 Map of United States' coal fields. (After Ref. 120, published by the U.S. Department of the Interior.)

that time, the production will slowly decrease over several hundreds of years (55). In the United States the yearly production of coal is now well over 500 million tons (101). If all of this coal would be put into railroad cars, the train would twice encircle the Earth. This is expected to rise to about 590 million tons by 1980 and 680 million tons by the year 2000 (118). From the early 1970s until about 2200, it is estimated that about 185 billion tons of coal will be mined, which is well within the estimated United States reserves (84).

There are about 1200 coal-mining companies in the United States. The ten largest companies account for about half of our yearly coal production. Several of these are owned by oil and gas companies, only one of the "Big Ten" (North American Coal Company) is independent. Oil companies control about 30% of the country's coal reserves. The reason for this is economical. Oil companies want to become energy companies and the world's longest lasting fossil energy source is coal. They also possess ample capital to develop mines and other facilities. Each new deep mine requires from 25 to 50 million dollars to build. Several major coal producers are owned by metal manufacturers who use metallurgical grade coal in smelters and blast furnaces. Metallurgical grade coal is in short supply in the United States and abroad; therefore, its price is high. The 1973 oil crisis caused a shift back to coal-fired boilers by electric utilities, driving coal prices to unprecedented heights. From the 1970 price of $5.89 per ton, utilities coal cost has risen to $18.69 in 1974. It is expected to rise to about $50 per ton, near the "per BTU" price of oil—the main competitor.

Due to recent legislation relating to mine safety, hundreds of small mines were shut down and many small companies were forced out of business. As a result, steam-generating coals were also in short supply when industrial firms and utilities contemplated a switch back to coal. It takes seven to eight years to get a new deep mine into production, even strip mines require two years. Mines that were abandoned earlier are not safe to enter, and they can hardly be put into production again if the need arises. Increasing coal production on short notice is hindered by shortages of trained miners, lack of mining equipment, shortages in explosives, fuel oil, roof bolts to hold up the mine's ceiling, and transportation equipment. Penn Central railroad reported in the fall of 1973 that it is 200 to 300 coal cars short on an average day. These experiences indicate that shifting from one energy resource to another may take great efforts and years to accomplish, even if the resource is abundant.

There is plenty of coal in other parts of the world as well. For example, the Ruhr area of the Federal Republic of Germany, where coal is mined from depths of 6000 feet, has reserves of about 10 billion metric tons, enough for over 120 years at the present rate of production. Coal seams located even deeper are estimated to have an additional 200 billion tons of reserve.

The two major cost factors in the utilization of coal as an energy source are its mining and its transportation (90). Most European and Russian coal fields are located deep in the ground where its mining is costly and hazardous (26). The layout of a typical deep coal mine in shown in Fig. 4.2. Mechanical mining causes dust and sometimes inadvertent sparks cause explosions. Coal mining in the USSR proceeds with hydraulic tunneling, using high-pressure jets of water to fracture the coal seam. This reduces the hazards but causes increasing water pollution. The dusty air of the mines has incapacitated some 50,000 miners in the United States. Mine operators and unions arranged pension plans for these miners, but the operators had no substantial responsibility to individual casualties other than through pension contributions.

In the early 1960s 34% of America's coal production came from surface strip mines. This percentage rose to almost 50% by 1974. Unfortunately, less than 10% of the US coal lies within 150 feet of the ground, a depth that limits the use of surface mining technology in the foreseeable future. The biggest earth mover, called the Gem of Egypt is a 14-stories-high device that moves under its own power at a speed of 0.25 mile per hour. It is shown in Fig. 4.3. It scoops up 103 cubic yards of earth at a time, operating on the Eastern Ohio coal fields of the Consolidation Coal Company. The cost of this machine was seven million dollars. Deep mining equipment is less productive and more costly. Modern mine machinery is shown in Fig. 4.4. New technology, such as the mechanized longwall mining shears a 1500 to 5000 feet long seam in several feet of depth with one pass, dumping the coal on a conveyor belt. Protected by movable steel supports, this system, if successfully introduced, will produce coal at a greater efficiency. Recent technological studies in Great Britain indicate that totally mechanized mining may become feasible in the future. The cost of mining, however, keeps increasing. While mining one million BTU worth of coal in 1969 was only 18 cents, in 1972 it rose to nearly 30 cents. Much of this cost increase is caused by new laws of land reclamation due to recent interest in ecological conservation.

Strip mining is economical—total delivered economic value of an 18-inch-thick seam is $12,000 to $25,000 per acre or $4 to $8 per ton. Usual American land reclamation costs of $1500 to $2000 per acre increases the cost of coal by about 50 cents a ton. Comparing this to the German practice

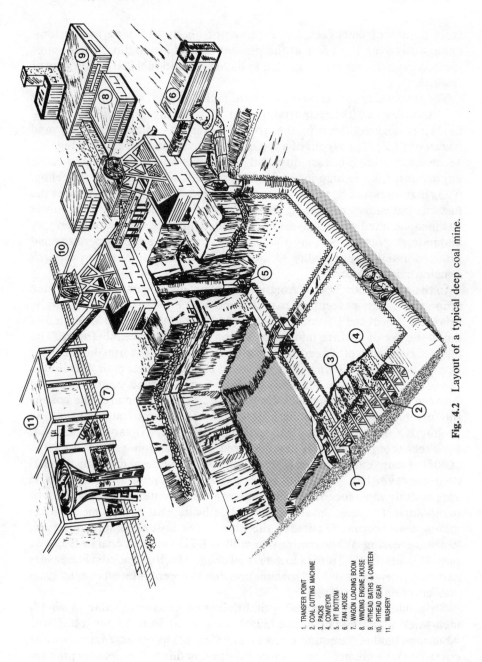

1. TRANSFER POINT
2. COAL CUTTING MACHINE
3. PACKS
4. CONVEYOR
5. PIT BOTTOM
6. FAN HOUSE
7. WAGON LOADING BOOM
8. WINDING ENGINE HOUSE
9. PITHEAD BATHS & CANTEEN
10. PITHEAD GEAR
11. WASHERY

Fig. 4.2 Layout of a typical deep coal mine.

Fig. 4.3 The world's biggest strip mining machine. (Courtesy of Consolidation Coal Company, Morgantown, W.Va.)

of full reclamation costing $8000 per acre, one may conclude that American strip mining companies are getting a good deal at the cost of despoiling the American landscape.

Today an area larger than the state of Connecticut shows the destructive scars of strip mining in the United States. Strip mining is economical even when a 20-foot layer of overburden has to be moved to reach a one-foot layer of coal. Coal companies own mineral rights to millions of acres of underdeveloped land in the eastern United States and are aggressively stripping its overburden, dumping it in unsightly spoils to extract the coal seams. The result is that large tracts in the Appalachian region are turned into wastelands with little hope of ever returning them to natural landscapes. The mining cuts cause additional landslides, creating further damage for which mine companies are not responsible by law.

Coal contains up to 5% of sulfur in addition to about 36 other chemicals

Fig. 4.4 Modern deep mining machine, the Marietta Drum Miner. (Courtesy of National Mine Service Co., Pittsburgh, Pa.)

that may act as pollutants. Water pollution through acidic mine water entering surface steams from mine openings destroyed countless streams in the Appalachian area. The sulfur content of these coals when exposed to the atmosphere causes yellow-gray acid to drain into the streams, killing life and plants in the water and causing corrosion to bridges and other structures. Water supplies of nearby cities and villages are also endangered.

Transportation of coal is commonly done in river barges or in special railroad cars each holding 50 tons of coal. Coal trains of 120 such "hoppers" are a common sight in our coal-producing regions. The process of grinding up coal and transporting it with water in slurry pipelines has been used in some places. The Cleveland Illuminating Company had such a pipeline in operation during the 1950s and a similar plan was advanced repeatedly to move West Virginian coal to the East Coast markets. Such systems are in operation in Australia and other parts of the world, but pumping and maintenance costs do not warrant building such pipelines where rails are already present. The transportation of coal energy through gas pipelines after gasification holds considerable promise for the future.

Coal's major problem is its waste products. In the United States about half of the coal goes for electric power generation. For that, the coal is ground into very fine powder by pulverizers, then blown into the burning space as shown in Fig. 4.5. The heat is transferred to water in a huge structure called a boiler. A typical boiler's cross-sectional drawing is shown in Fig. 4.6. The steam generated powers a steam turbine, which in turn rotates an electric generator. The smoke and gas pass through the stack and contribute to air pollution. Electrostatic precipitators—a costly proposition—remove nearly all the mass of fly ash, but they pass the invisibly small particles that are about 0.001 millimeter in size. These cause respiratory problems. The submicron particles (less than 1/22,000th of an inch) make up much of the smoke from power plants. A new steam-scrubbing method, developed by Lone Star Steel Company of Texas, reduces these emissions from 40 tons to 40 pounds a day.

The sulfur content of coal, up to 5%, enters into the atmosphere as

Fig. 4.5 Coal powder burning in the fire chamber of a boiler. (Courtesy of Babcock & Wilcox Co., Barberton, Ohio.)

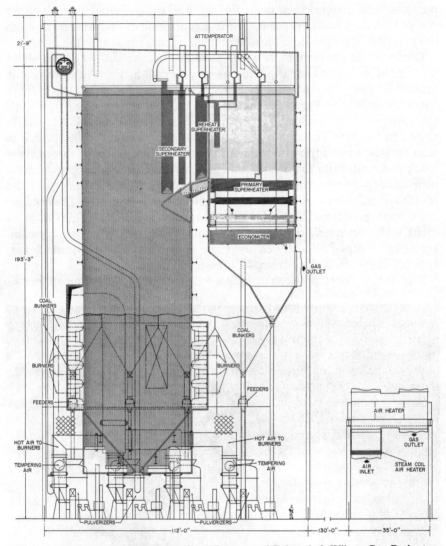

Fig. 4.6   Cross section of a steam boiler. (Courtesy of Babcock & Wilcox Co., Barberton, Ohio.)

sulphur dioxide, a highly corroding compound that cannot be removed from stack gases by any existing technology. Governmental regulations require that the sulfur dioxide emissions be below 0.3% to 0.5% from industrial power plants. This prevents the use of most of our eastern coal fields. Nitrogen oxides form, causing further air polluting gases. It is said that such air pollution costs a minimum of $10 billion to the United States annually in corrosion, cleaning bills, etc. The value of lost human life due to cancer and other diseases attributed to air pollution cannot be enumerated. It is known that the urban rate of cancer exceeds that of the rural data by a factor of two or three. The distribution of cases of multiple sclerosis corresponds well with the regions of high air pollution. These considerations have led to restrictions on sulfur content in fuels, especially along the eastern part of the United States. In effect, these are restrictions on burning coal because the eastern coals are usually high and the western coals are usually low in sulfur content. The coal market generally exists on the eastern part of the United States.

Building power-generating stations away from populated areas does not solve the problem either since it only tends to transpose the point of origin of the air pollution without eliminating it.

A recent AEC-sponsored study conducted by the University of Utah indicated that a radioactive gas—radon—is released into the atmosphere when fossil fuels, coal and natural gas, are burned. Accordingly, burning these fossil fuels is apt to increase hazardous radioactivity in the air and it could conceivably produce a greater annual lung dosage of radioactivity than the fallout during the peak of atmospheric testing of nuclear weapons.

Several new technical proposals are under consideration to reduce the polluting characteristics of coal. In Europe research is in progress to build coal burners where the coal, instead of being pulverized and blown into the fire chamber, is transported into the fire on a movable grate with air added from below. This method reduces the fly ash content of the smoke and presents different ways for keeping sulfur oxides out of the flue gases. A $3.5 million investment by General Motors resulted in the development of a process that removes 90% of the sulfur dioxide from the smokestacks. First the noxious gases are "washed" in a chemical and water solution. This polluted water is then purified and recirculated through the system. The purification involves adding lime to the liquid, which results in a calcium sulfate material much like a poor grade of wallboard that is disposed of in landfills.

The Japanese Mitsui scrubber system was also found highly successful in operation. Cost of air pollution control systems for electric generating plants burning coals of high sulfur content are very expensive. As much as $200 million may be spent on a single plant for scrubbers. Maintenance of these systems is complicated and its added space need is heavy. Also, the continuous need for lime and soda ash in scrubber operations could strain the production capability of these materials if widely introduced. One scrubber uses about 6,000 tons of lime and several thousand tons of soda ash each year. In all, the elimination of sulfur dioxide emissions from industrial and utility companies' smokestacks could add to the price of industrial products and electricity significantly.

Other new methods of combustion, in which the pollutants are bled off before they reach the atmosphere, are in the research stage. Of course, special new combustion chambers are necessary for these methods, so this new technology can rarely be utilized with existing power-generating plants. But the idea to remove pollutants before they reach the smokestack is a logical one that deserves the research effort toward its solution.

To make high sulfur-containing coal useful under the Clean Air Act, it would be advantageous to develop a new process to remove pyritic sulfur from some eastern coals. This technique was demonstrated to be chemically feasible at the pilot process level. The cost does not seem to add considerably to the processed coal price. However, most coals (as many as 50%) seem to contain sulfur in a chemically bound form rather than in separate crystals of pyrite. A promising process to treat these coals is solvent refining. In this process coal is first dissolved in a reagent, which then is treated chemically to precipitate the sulfur. A privately built pilot plant was completed in 1973, another with federal sponsorship will be on line in 1975. There are several other coal liquefaction processes that shown promise but no commercial demonstration-size plants are yet planned.

A 1971 study in Germany indicated that manufacturing gasoline from coal using a synthesis process would be possible but it would cost 72 cents per gallon. The older hydration process, developed in Germany in 1936, would produce gasoline at a cost of 58 cents per gallon. By modernizing this process, further savings may be accrued. These, and the ever-inflating cost of the natural crude oil, may make it feasible to produce gasoline from coal, which would be competitive with refined oil.

Gasification of coal is given a great deal of attention by the United States government, particularly since the 1973 oil crisis. With improving technology, the gasification of low quality coals is expected in the future; this will increase efficient coal utilization and, therefore, the mining of coal (27). In 1973 the Continental Coal Company reported success in the experimental gasification of West Virginia coals of high sulfur content.

Coal gasification—that is, manufacturing a combustible gas from coal—involves the fact that coal consists of about ten parts carbon and eight parts hydrogen, both burnable elements. The other main element is one part oxygen. To form coal into a burnable gas, we would have to find a way to add more hydrogen to this compound. Water is by far the most abundant source of hydrogen. Presently hydrogen is produced from natural gas, which would be a backward source for coal gasification. If a suitable technique could be found to improve the chemical reaction of carbon with water one could could produce an ideal fuel in the form of carbon dioxide and hydrogen. Such a method being nonexistent at present, the serious coal gasification proposals use different intermediate reactions to overcome the troubles. There are two main schemes. One method produces low-energy gas (200 BTU/ft^3, 8 megajoules per cubic meter) in large volumes as the process uses air as an oxygen source, leaving air's nitrogen gas in the product. This resembles the old-fashioned "city gas," from the days when gas lights were fed by gas manufactured from coal. The main advantage of this process is that about 90% of the coal's original energy remains in the product, but this new energy appears with the hot gas in the form of heat. Cooling it would obviously be a waste of energy; therefore, this type of product can best be converted into electric energy immediately by placing the generating plant in the mining area and transporting the electric current by wire. This would eliminate the considerable transportation difficulty inherent in moving the large volumes of low-energy gas by pipelines.

Another coal gasification method produces high-energy gas, mainly methane with virtually no inert constituents. The product contains about 39 megajoules/m^3 and is almost identical to natural gas. Therefore, it can be shipped by our existing national pipeline network. The disadvantage of this scheme is that about one-third of the coal's original energy is lost as waste heat of the conversion plant. Another disadvantage of the process is its high cost, amounting to about $20 per ton of coal. It is, however, expected that further development will reduce the cost and convenience; cleanliness of the product makes the method worth the effort. A $7 million pilot plant built in Chicago is now in operation and two more are coming along.

The latest gasification technique is the Hydrane process. Still in the early stage of development, it consists of two steps. In the first step, called the free-fall phase, pulverized coal is fed into the top of large reactors. As the coal falls through the reactor, a mixture of hydrogen and methane—called the feed-gas—is pumped in.

This mixture reacts with the coal, converting about 20% of its carbon to methane—the primary ingredient in natural gas. The gas coming out of the reactor is about 70% methane. This gas then undergoes a small amount of methanation to produce even more methane. The final product has a heating value fairly close to that of natural gas, and is ready to be put in pipelines.

In the second step, unconverted coal from the free-fall reactors passes into a dilute-phase reactor. Here, a mixture of hydrogen and methane is produced, and this becomes the feedgas to the first part of the system. In this stage, an additional 20% to 25% of the carbon is converted, meaning that overall about 45% of the carbon is converted to methane, and we still have carbon left in the char that comes out of the dilute-phase reactor. This carbon is used to produce the hydrogen for the process.

The Hydrane process has a number of advantages over other coal gasification methods under development. The small amount of methanation needed allows conventional equipment to be used, rather than the still-unproven equipment needed for extensive methanation. The overall system would be simple in design, meaning that there would be fewer shutdowns for repair. Because it uses hydrogen to convert most of the coal, the system would eliminate most water purification problems. But perhaps the greatest advantage is its high thermal efficiency.

The American Gas Association has already made a study to determine where gasification plants could be located. The findings indicate that there are 176 sites that could support an average-sized plant. About 50 of these are east of the Mississippi River, and the rest are in the West. Each of these plants could produce about 250 million standard cubic feet of gas per day—about one-half of one percent of our current daily consumption. Still federal funding of coal gasification research until 1974 was relatively scarce. The United States Bureau of Mines' Office of Coal Research and the Atomic Energy Commission are pursuing the problem along with private industry. It is expected that by about 1980 commercial production of gasified coal will commence. A plant producing 250 million cubic feet of gas a day would cost about $300 million and would use 16,000 tons of coal daily. Consolidation Coal Company has purchased 200 million tons of coal reserves in Pennsylvania and West Virginia for gasification when the process will become financially feasible. It is expected that within a generation gasified coal will take the place of our dwindling resources of natural gas. The result, though, will be costly. Gas from the proposed Pacific Lighting Plant in California will cost $1.20 to $1.30 per thousand cubic foot, compared with an average of $0.40 per thousand cubic foot for the company's current supplies of natural gas.

Research on coal gasification using nuclear reactors was recently announced by American firms. Although it may take many years to bring this process to production capability, it may be a significant step toward solving the gas supply problem.

A promising new technology to produce methane gas by burning coal *in situ* underground is also in progress. This method, if successfully developed, would eliminate much of our present difficulty in coal utilization.

Although several coal gasification processes are under investigation, none of them was proven to be fully economical and technologically reliable. They also have potential environmental problems. The disposal of the ash residue, the prevention of atmospheric emissions, the control of coal dust, water pollution and similar problems are yet to be worked on. Even if all of these problems are quickly solved, experts claim that gas from coal will not take a major share in America's energy supply before 1985.

CHAPTER 5
Oil

A major share of the current energy supplies of the United States comes from petroleum. It is, like coal, of organic origin, developed during long periods of geologic history, at specific locations where physical and chemical conditions were favorable. It is therefore an energy resource of stored solar power of ages long gone; once found, extracted, and burned, it is lost forever. The energy content of crude oil is 5,800,000 BTU per barrel. One barrel (the measure of oil) contains 42 US gallons of the liquid.

Petroleum is a thick, viscous liquid of brown-dark green or, sometimes, light color with a characteristic smell. Basically, it is made up of carbon and hydrogen. Small amounts of sulfur, nitrogen, and oxygen are also common in most oils. The carbon-hydrogen compounds appearing in crude oil are in the thousands. Depending on the formation history of the oil, certain types (or series) of hydrocarbons are usually predominant in certain areas. Oils of the Pennsylvania area are generally of the paraffin series, oils of Texas and Baku (USSR) are of the naphthene series. In the paraffin oils there is a higher hydrogen content in relation to carbon, in the naphthene oils this relation is reversed. Naphthene oils are heavier and are richer in viscous lubricating oils. Distillation of naphthene oils results in significant amounts of solid or semisolid asphaltic residues. Oils of Mexico are rich in sulfur (5%), while Pennsylvania oils contain little sulfur (0.5%). In spite of the variation of carbon-hydrogen compounds in oils the carbon-hydrogen elements percentage distribution is surprisingly constant in most oils. They contain about 85% carbon and 13% hydrogen. When warmed in a distillation column, the mixture of hydrocarbon compounds separate according to their boiling points. First to evaporate is gasoline, when warmed to 220°C; warmed further to 280°C, kerosene is separated; further warming to 350°C produces diesel oil. The remaining heavy hydrocarbons are vacuum-distilled to produce lubricating oils. Further distillation produces asphalt and graphite. The products of the distillation are usually further refined by chemical treatments. The technique of "cracking"— breaking down long-chain, heavy hydrocarbons into short-chain, gasoline types—is accomplished by high pressure and temperature in the presence of catalysts. This method allows the control of the proportion of the fractionated petroleum products, regardless of the type of the original crude.

Distillation, cracking, and other refining processes are made in pet-roleum refineries. In the early 1970s there were about 250 refineries in operation in the United States. About 58% of the American refinery capacity is held at eight major oil companies. These are Arco, Exxon, Gulf, Mobil, Texaco, Shell, Standard Oil of California, and Standard Oil of Indiana. The pivotal point of the current gasoline shortage is that of inadequate refinery capacity. It costs about $200–250 million to build one new refinery producing 150,000 barrels per day. Due to unfavorable taxation, government regulations, scarcity of capital, and vigorous en-vironmental opposition, the construction of new refineries is mostly blocked. Since the discovery of petroleum, 1972 and 1973 were the first years without a new refinery being out into operation. Meanwhile, the nation needs about five or six new refineries of 130,000 to 150,000 barrels a day capacity every year through the next 15 years in order to solve the current fuel crisis in the United States (130). Until then, oil imports from abroad will, in part, come in the form of foreign crude refined in European or Virgin Island refineries. It is rather unlikely that the schedule mentioned above will be met. Between 1969 and 1974 there were 1.9 million barrels a day added to American refinery capacity. In the same period demand for petroleum products climbed 3 million barrels a day. The welcomed change in oil import regulations resulted in announcements of new refinery constructions by American oil companies, with a projected combined capacity of 4.5 million barrels a day. Even if all these commence, relief is not in sight for the near future. It takes at least at least two to three years to complete a new refinery. In the early days of 1974 the US refining capacity stood at 14.1 million barrels a day. The daily consumption, in comparison, was about 17 million barrels. Imports made up the difference. This situation has developed during the past few years. In 1965 the gap between domestic refinery capacity and consumption was only 0.5 million barrel per day. Before the 1973 oil crisis, the growth rate of American oil consumption was about 5% to 6% yearly. Due to conservation measures, this growth rate was reduced to about 2.5% in 1974. Projected increases of American refinery capacity will not be able to close the gap even with this small rate of growth in demand. Legislative threats of price rollbacks and heavier taxation of oil companies could potentially worsen the situation in the future. The problem is most critical on the East Coast. Forty percent of our petroleum products is used up there, yet only 15% of the refinery capacity is located there. Eight major oil companies abandoned plans to build new refineries along the East Coast because of restrictions against refining high sulfur crude oil. Low sulfur crude is in extremely short supply around the world. These and similar environmental considerations work against the easing of the energy problem.

Before we go any further, let's see how crude oil came about. Why is it relatively rare on Earth? Why couldn't we just dig up some more to solve our problems? Formation of crude oil requires bays, gulfs, coastal lagoons, and enclosed seas like the Black Sea or Caspian Sea, where ocean currents cannot enter and thus the lower layers are motionless. The lack of mixing of water prevents oxygen from the air entering the bottom layers. Under these conditions, dead organic matter accumulates on the bottom and, with the aid of anaerobic bacteria (those that do not require oxygen), the formation of hydrocarbons begins.

Chemical and physicochemical analyses of petroleum show the presence of cholesterol, indicating that it is formed from the fats, waxes, and resins of marine animals, planktons, and vegetation (121). The optical properties of petroleum present proof that it is mainly of animal origin, while coal is formed mainly from plants. Due to the bacterial action, there is more hydrogen in petroleum than in coal.

Clay, silt, and limestone deposited during the petroleum-forming process form the source rock (42). In order to retain the crude oil it is necessary that the bays, gulfs, or entrapped seas remain separated from the open seas (and hence oxygen), which would decompose the organic substance into carbon dioxide gas. The geologic process necessary for the formation of crude oil (and natural gas) is that the source rock be covered by a heavy impervious layer of sediments. This in most cases squeezed out the oil and gas from the source rock and caused it to migrate into nearby porous rocks like sands, gravels, and sandstones. Under the pressure of the heavier water in the pores beneath, the oil moves upward. If it is allowed to reach the surface, it oxidizes and the more volatile compounds escape into the atmosphere. Such outlets were known to our ancestors as sources of asphalt and naphtha, used by the ancients as lamp oil, medicine, or sealant. Asphalt was extensively used by the Babylonians in their dams and buildings as well as for mummification of the dead. (*Mumia*, mummy, means asphalt in Persian.) In the oil fields an impervious cover layer of a dome shape is usually present; this prevented the loss of the hydrocarbons (17). Prospecting for oil is an indirect process; geologists and geophysicists are using various prospecting methods on the Earth's surface to determine the location of geologic features of the underground rock layers and identify possible regions of the reservoir rocks where oil and gas might be trapped (49). These are the high portions of smoothly bent layers, slanting broken layers, or rock along the fault lines, and the like. Once such geologic features are located, exploratory bore holes are sunk into the layers. As oil is formed in the regions of receding shallow seas and gulfs, so are salt deposits. Often oil is found in areas where salt deposits are present. Salt water being heavier than petroleum, it is usually located below it. Gas, on the other hand, is lighter than oil, taking the place above it. In many cases all three are present in reservoir rocks. The oil, once found, in many cases will come to the surface, or near the surface, due to the pressure of the compressed gas about it or due to the pressure of the brine below it. The rate at which oil is allowed to flow out of the layer is controlled by many factors. Since both the gas and the water flow easier in the pores of the rock, if oil is pumped too fast it may cause the water to break through the oil layer to flow into the well. This ruins the productivity of the well by cutting off the oil flow. For this reason, well production rates are carefully controlled. Even with such controls, much of the oil (about 30% on the average) is left in the reservoir. To increase the flow rate, gas or water may be pumped back into the reservoir through old wells to increase the pressure in the reservoir. This may resurrect old fields such that their original production is multiplied manyfold. Private industry is learning to extract more fuel from existing fields. Secondary and tertiary extraction methods now being developed promise to substantially decrease the current 30% to 35% recovery loss, which is set by economic limitations.

To increase the flow of oil from the reservoir, the wells are sometimes treated by pumping in acid to dissolve the rock and increase the size of the flow channels in the cracks and pores of the rock. Blasting nitroglycerine in the producing layers has long been used also. Hydraulic fracturing—pumping in heavy incompressible liquids at high pressure to crack open the layers of reservoir rock—was introduced in the 1940s, increasing productivity by phenomenal rates. Some oil fields treated in such manner increased their yield manyfold. Other methods used to increase the flow of oil into the wells include warming up the oil around

In 1969 the oil depletion allowance was cut from 27.5% to 22%, depriving the petroleum industry of many hundreds of millions of dollars in capital funds annually.

Capital expenditure percentages of oil companies in the United States include

lease acquisition	17.3%
drilling and equipping exploratory wells	42.5%
drilling and equipping development wells	21%
oil and gas production and process equipment	5.5%
natural gas plants	5%
improved recovery programs	8.7%

As this tabulation shows, a large amount of oil capital goes into exploration for new sources (28).

In spite of rising costs of labor and materials, unfavorable taxation, and restrictive federal control, exploratory drilling goes on in the United States to increase domestic supplies. During the 1973 oil crisis, 1411 oil rigs were drilling in the United States and about 330 were at work in Canada. The all time high was in 1954, when 3100 rigs were at work drilling for oil and gas in the United States.

Among the most powerful enemies of domestic oil exploration are the environmental groups. Due to the disastrous oil leakage that occurred off Santa Barbara, California, offshore drilling suffered from understandable adverse publicity. On the other hand, the safety record of offshore drilling is remarkably good. Twenty-five years of experience in drilling for gas and oil off the coast of Louisiana and Texas, during which some 17,000 wells were drilled, shows that only 27 blowouts have occurred. This amounts to less than 0.2% of all offshore wells drilled. Of all these blowouts only the one at Santa Barbara caused significant contamination of beaches and shorelines. According to authoritative sources, permanent environmental damage at Santa Barbara was minimal.

In the experience of the Buzzard's Bay (Cape Cod) oil spill in 1970, the oil was found to persist in the marshlands for years, completely killing marine life when it washed ashore and affecting growth there for the duration of its stay. But at deeper parts of the sea bottom, concentrations of oil were found in the range of parts per billion. Toxicity is considered to be only in the range of parts per million.

Of the world's initial energy content, 3.25% is oil estimated to be somewhere between 2100×10^9 and 1350×10^9 barrels. This amounts to some 70 to 80 cubic miles of oil. This does not include the oil content of tar sands and oil shales—sources of some 490×10^9 barrels of oil. Distribution of the world's oil is estimated to be as follows (62):

United States	200×10^9 barrels
Canada	95×10^9 barrels
Latin America	225×10^9 barrels
Europe	20×10^9 barrels
Africa	250×10^9 barrels
Middle East	600×10^9 barrels
Far East	200×10^9 barrels
USSR and China	500×10^9 barrels

Other published data (129) on the world's oil resources show somewhat different figures. The discrepancy can be explained in two ways. One is that all of this information originates from the oil companies' own releases, a largely unsubstantiated and suspect source. The other is that "proven resources" are interpreted in oil trade lingo as the oil that can be produced and marketed at the current economic and technological conditions. Hence, every price rise means a large increase in "proven resources." New technological developments increasing the ultimate productivity of existing fields increase the "proven resources" as well.

In totalitarian countries domestic oil production is viewed as a means of survival in a political and military sense. Hence it is stressed against heavy economic odds in order to gain independence from uncertain foreign supplies. China's originally miniscule domestic oil production reached over 60 million tons in 1974 and is expected to hit 100 million tons by 1980. The Soviet production was less than half of that of the United States in 1962, by 1974, they produced about as much oil as the US and by 1980 they will produce significantly more than the U.S.

New oil-producing regions are continually being found. Substantial oil strikes were reported in 1973 from the Gulf of Thailand, the Malaysian section of the South China Sea, Newfoundland, Peru's Amazon river jungle, Venezuela's Orinoco river region, Indonesia's coastal areas in the Java Sea, China, and other sites around the world. Such discoveries will add considerably to the world's oil supply. From the standpoint of potential supplies for the United States, major oil finds reported in Mexico in October 1974 may be the most significant. Initial reports indicate that these new oil fields may boost Mexico's oil exports tenfold in the coming years, but at prices matching those of the Arab countries.

The list above shows that the world's major suppliers of oil are the Arab countries (130). As an example of the rate of use of these reserves, daily oil production of the Arab countries in 1972 and expected production in 1980 are shown below (44):

	Million barrels per day	
	1972	1980
Oman	0.3	0.5
Qatar	0.4	2.5
Abu Dhabi	1.1	3.0
Iraq	1.1	3.0
Libya	2.0	2.0
Kuwait	3.0	3.0
Iran	5.2	8.3
Saudi Arabia	5.2	10.0
TOTAL	18.3	32.3

In contrast to these figures is the fact that in 1973 the United States used about 17 million barrels of oil daily, while its production was considerably less than that. About 25% to 30% of this amount is supplied from foreign countries. In 1973 about $8 billion worth of foreign oil was imported into the United States; this amount was expected to rise to $17 billion by 1980. Our 'Operation Independence" may change this picture in the future.

As the American consumers' oil requirements expand by about 4% each year, it is estimated that about 26 million barrels per day will be used by 1985 in the United States.

Since the turn of the century, the United States has been the largest oil producer of the world, followed by the Soviet Union, Iran, and Saudi Arabia. The following list shows the production of some major producers in the year 1971.

United States	490 million tons
USSR	377 million tons
Iran	227 million tons
Saudi Arabia	223 million tons
Venezuela	147 million tons
Libya	132 million tons

Since November 1971, when the daily US production of crude oil hit the 10.8 million barrel mark, the output has declined to about 8.5 million barrels a day by May 1974. In the same month Saudi Arabia's oil fields produced about 9 million barrels a day taking away America's leading position in oil production. Experts indicate that the Soviet Union's oil production will soon top that of both Saudi Arabia and the United States.

It is estimated that oil production in our country will be 14 million barrels per day in 1985 and 10–18 million barrels per day by the year 2000. Therefore, America will be depending on foreign oil in the future (108), unless steps are taken to avoid this.

Crude oil and petroleum products imported from major foreign sources into the United States in 1970 were as follows (35):

Source	Crude Oil	Refined Products	Total
	(in million barrels)		
Canada	245.26	34.47	279.73
Venezuala	106.11	641.55	747.66
Dutch Antilles	—	174.19	174.19
Puerto Rico and Virgin Islands	—	99.09	99.09
Europe	—	64.48	64.48
Iran	12.18	1.85	14.03
Saudi Arabia	14.54	0.52	15.06
Libya	17.16	0.10	17.26
Algeria	2.09	0.98	3.07
Nigeria	17.49	0.63	18.12
Indonesia	25.67	—	25.67

In terms of energy, the United States has imported a rapidly increasing amount from foreign sources in the past few decades. This, in 1970, amounted to 5679 trillion BTUs (mostly in the form of oil), about 8.4% of the total energy used in the United States. This amount has risen to 7046 trillion BTUs by 1971, comprising 10.2% of our total energy used in that year. By 1974 one-third of our oil supply came from abroad.

America possesses 7% of the world's oil reserves, while using one-third of the world's oil production (31). The tabulations above allow a glimpse of the fact that the most technically developed areas of the world—the United States, Europe, and Japan—use an overwhelming share of the world's oil.

Of foreign countries, Japan is the largest importer of oil, followed by Italy and Great Britain. The list over shows the 1971 data of major foreign oil imports:

Japan	200 million tons*
Italy	116 million tons
Great Britain	110 million tons
France	108 million tons
Germany	100 million tons

The producers of this oil, mainly the Arab countries, collect a phenomenal price for their oil, which will amount to a quarter-trillion dollars or more between 1973 and 1980. The dislocation of these funds is supected of being one of the major factors in the world's monetary crisis in this period (11). The price of foreign oil has quadrupled between 1970 and 1974. Seventy-nine percent of our hydrocarbon fuels are used to heat our homes, power our automobiles, etc. The remaining 21% is used to generate electricity or to supply new material for many industries (plastics, etc.). It was estimated that by 1985, at the present growth of gasoline use due to emission controls, etc., US consumption of oil may rise to 24 million barrels daily. Without substantial increase in domestic production, more than 60% of our demand must be met from foreign sources. This would create an intolerable balance of trade deficit of $25–30 billion per year. Although this would be reduced by exports, the remainder would still be huge.

In the recent past there have been some spectacular finds of oil supplies in the world. Most important from an American standpoint was the location of Alaska's oil fields in the Prudhoe Bay region. Early in 1968, exploratory drilling on the North Slope of Alaska confirmed the existence of about 10 billion barrels of recoverable oil, increasing United States oil reserves by 35%. Speculative estimates put the eventual Alaska oil reserve to be somewhere between 30–50 billion barrels. It must be kept in mind, however, that 50 billion barrels would be less than a ten-year supply for the United States at the present rate of consumption.

Another recent significant oil find is that of the North Sea, which is estimated to exceed the capacity of Alaska's North Slope. The significance of this new oil source is great from the standpoint of the European continent. It lies a little over 100 miles off the Scottish coast, while Persian Gulf oil must travel 11,000 miles around Africa. Oil from the North Sea will come on line in 1974, and is expected to supply about 15% of Europe's energy needs by 1980. By the mid-1980s Britain is expected to fill all her needs from her North Sea sources and will become an exporter of oil. The technological challenge to bring this oil out is formidable. One

*One metric ton of oil equals 7.4 barrels.

hundred mile per hour winds and 90 foot high waves require capital expenditures for drilling that are 20 times higher than those encountered on dry land.

Perhaps the largest new find of the world was reported by the USSR from above the Arctic Circle in Siberia (120). East of the Ural Mountains at Samotlor, fields appear to be holding the world's biggest single oil deposit. Drilling in temperatures in which brake fluids freeze and steel becomes brittle, 3000 oil wells were built at a rapid rate. After the completion of this field, Soviet export of oil is expected to rise from 70 million metric tons in the early 1970s to levels where the USSR may be considered to be one of the world's major oil exporters.

Another potential new field of interest to the United States is that of George's Bank off the New England Coast. Perhaps up to 10 billion barrels of oil may be extracted from this area of 14,000 square miles, if environmental objections to offshore production are overcome. Another promising site is the Baltimore Canyon Trough about 12,000 square miles off New York, New Jersey, Delaware, and Maryland, where some of the thickest—and possibly most favorable—geologic formation for oil occurs. Still another site is the Blake Plateau, more than 70,000 square miles, stretching offshore from Cape Hatteras south to Florida. The plateau, a gently sloping portion of the continental shelf, is the largest prospective area for oil and gas—but also the most difficult because it lies in deep water as much as 200 miles offshore. Our continental shelf probably contains 160–190 billion barrels of oil, 25–30 billion barrels of natural gas liquids and 200–1100 trillion cubic feet of natural gas that are recoverable. Our potential offshore oil fields are shown in Fig. 5.2.

In spite of these new finds it is expected that increasing demand for crude oil in the world will use up most of the available reserves within the next generation. Based on estimates of the world's total petroleum reserves and the rate at which they are being used, the probable cycle of the world's production will reach its peak before the year 2000. By that time the yearly production of oil will perhaps be double the production of the early 1970s. But after another 25 years, in about 2025, the production will fall back to today's level and will rapidly dwindle afterwards as oil resources run out (77).

In essence the supply will start to dwindle from a peak before the year 2000. Therefore, is necessary that new technology be developed to secure synthetic oil from coal beyond that time or new sources of energy must be found. But significant major replacements for petroleum are not expected in the near future. Meanwhile the movement of oil from producer to user

Fig. 5.2   Map showing offshore oil fields on the United States East Coast.

continues in an expanding rate. Transportation of petroleum is a major factor in its price (103). This movement is tremendous (75). Some figures of transported oil in 1970 are shown below:

Arabic peninsula to Europe around Africa	5.004 million barrels per day
Arabic peninsula to Japan	5.11 million barrels per day
Libya to Europe	4.6 million barrels per day
Venezuela to overseas	3.632 million barrels per day

The daily transportation of these amounts is a considerable technical problem. First the crude oil should be transported by pipelines, truck, or rail to terminals. Storage of the oil at the fields as well as at the terminals requires considerable volumes of storage tanks. Storing oil from offshore fields is even more costly. On the North Sea oil field, Phillips Petroleum had a Norwegian firm construct a 42-million gallon capacity double-walled storage tank to float to sea. The outer wall of this tank is perforated to absorb the impact of the giant waves common at that location.

To fill the tankers that carry the oil across the oceans, large specialized port facilities are required. Due to the size of the most modern ocean-going tankers, only a few American ports can handle oil shipments in the future. Plans are being considered to construct offshore floating platforms to handle the oil traffic. The technical problems and ecological dangers are considerable in building and operating these proposed facilities.

Oil tankers of normal size are already expensive. Standard supertankers of 100,000 tons cost about $30 million in 1973. About six 200,000 ton carriers would be needed along with port facilities to supply 9 million barrels a day. Present United States port facilities can handle 20 of the largest allowable tankers a day. The use of very large crude carriers of 250,000 to 400,000 deadweight tons is desirable because they provide cheaper transportation and less danger of spills.* Proposed million ton giant supertankers will probably reach the limit of feasibility in the near future. Today there are no United States ports to handle such ships. Without such port facilities, by 1985 imports of estimated 13.5 million barrels per day would require the daily unloading of about forty 50,000 deadweight ton tankers along the nation's coastline. Deep water terminals must be built in order that the larger carriers' benefit be gained. Ports like Delaware Bay should be dredged to a depth of 90 feet to handle modern tankers.

By 1980, the world's oil tanker fleet should be about 450 million deadweight tons; 312 million deadweight tons are yet to be built at a cost of about $47 billion. So it seems that moving oil around by ships is an expensive proposition.

Undersea pipelines may carry oil to relatively short distances; ocean-crossing pipelines will probably never become feasible. Some shorter distances, such as the almost 200-mile distance from Teesside on the English coast to the Norwegian offshore oil field and an over 100-mile long distance from Aberdeen, Scotland, to the British Forties field will be constructed in the near future, although the danger of a pipeline break under the sea and resulting ecological catastrophe is rather high.

Perhaps one of the most typical problems of oil transportation is that of Alaska's North slope fields (97). First a proposal was made to build a 48-inch pipeline through Alaska to the Port of Valdez, where the oil would be transferred to the United States' West Coast. The daily capacity of the pipeline would ultimately be two million barrels.

Another proposal was to carry the oil on special ice-breaking tankers

---

*The world's largest tanker in 1973 was the Globtick Tokyo, 477,000 deadweight tons.

through the Northwest Passage. To test this idea, the Humble Oil Company converted the tanker Manhattan, which made an experimental 4800-mile trip to test the idea. Due to various problems encountered, this idea was shelved.

The third plan to deliver Alaska's oil to the continental United States was to construct the 3200-mile trans-Canada pipeline running east of the Rocky Mountains through the Mackenzie Valley to Alberta.

The fourth idea was to build heavy duty, two-track railways from Montana to Alaska. This railroad could either be built through the Mackenzie Valley or through the Yukon Valley.

The fifth idea (proposed by Boeing Company) was to build mammoth tanker airplanes with 12 jet engines each. These airplanes would have an incredible 355 million pounds of gross weight at takeoff and they would carry 8100 barrels of oil each trip at a speed of 460 miles per hour. Thirty-seven such super tankers operating 20 hours per day on flights of 575 miles to the nearest ice-free port could deliver two million barrels of oil daily. The cost of one of these planes would be in excess of $70 million.

Apparently the interested oil companies were committed to the first pipeline proposal, about 800 miles of pipe have already been purchased for that by 1973. Indeed, it appeared the most feasible proposal. Environmental dangers cited by opponents to the proposed pipeline have temporarily stilled its construction. Meanwhile Alaska's oil rests in the ground. Opposing ecologists say that the hot oil would melt the permanently frozen land and would obstruct the natural paths of migrating caribou herds, risk breaking along Alaska's earthquake belt, and raise the prospects of oil spills in the stormy North Pacific from the tankers that would carry the oil from Valdez to the United States coast. The Alaska pipeline was approved by the United States Congress by the end of 1973.

In the light of international difficulties making our oil imports uncertain (Middle East shipments were interrupted six times since 1950) domestic oil production is certain to be increased in the near future. One of our most important oil resources in the United States is the content of the oil shales covering large areas in the western states as shown in Fig. 5.3. In addition to the rich northwestern Colorado, Utah and Wyoming also have considerable deposits.

Oil shale is a sedimentary rock deposited in layers. It can be mined either on the surface or underground. The oil is removed from the rock by technological means, but in the process the waste rock expands in volume by at least 12%, causing disposal problems. The surface mining would

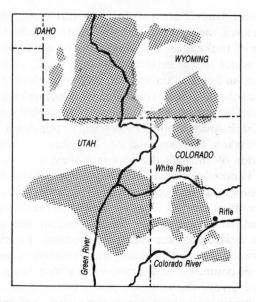

Fig. 5.3 Map of United States' oil shale deposits.

result in permanent scarring of the land, much like strip mining of coal. Objections of environmentalists as well as most local citizens will probably force the producers to commit large funds for reclamation purposes.

The world's oil shale resources account for about 3100 billion barrels of which about 190×10^9 barrels are considered recoverable by today's technology. This is about 0.32% of the world's initial energy resource. Much of this is located in the United States. Oil shales are found also in Mexico, Syria, Argentina, and in the Turkestan region of the USSR. Shale oil has been produced economically in Scotland for many years. The US Bureau of Mines opened an experimental plant to produce oil from western shales back in 1926. Unfortunately, this experiment was abandoned within a year due to lack of money. In 1945 another such plant was opened, only to be closed a few years later. The American oil shales contain 25 to 30 gallons of oil per ton of rock. The oil is of low sulfur content, hence it is "clean fuel." Mining of oil shale in Colorado began, on a prototype basis, in the summer of 1973. Initially, less than 250,000 barrels a day is to be produced. By the 1980s the production will reach a million barrels a day. This will be a large industry, requiring mining, transportation, processing, and storage facilities (112).

By bringing the United States oil shales to the market, the American energy reserves will increase by 3.5×10^{18} BTU, for the better quality (25 gallons per ton of rock) shale. Low-quality oil shales are more abundant. These contain about 10 gallons of oil per ton of rock. There are enough of these rocks to add a further 12×10^{18} BTUs to our reserves. These resources are enormous when compared to either our oil or gas reserves in this country. The magnitude of shale oil deposits in our country make even the Middle Eastern oil fields small in comparison. Their extraction is, therefore, a prime national interest.

The production of oil from shale is being studied by laboratories. New technology will undoubtedly decrease the present high cost of production. Present technology calls for conventional mining of the shale, followed by heating it in retorts to 950°F to melt out the oil. This oil would then be refined as usual. New methods, like NASA's research in using powerful laser beams to disintegrate the rock, may become feasible for industrial use by the 1980s. A new way of getting the oil out was devised recently by an American oil company. This involves cracking the rock deep in the ground by explosives. Then natural gas is pumped into the cavity and fired. The resulting heat would allow the oil to flow out through the cracks and collect in the cavity, where it then could be pumped out in conventional ways. The cost of this oil would amount to no more than $2 or $3 per barrel.

Another source of oil is the content of tar sands found in abundance in Canada's western parts (Northern Alberta). The tar sands contain even more oil than oil shales. The world's tar sand oil resource is said to be about 300×10^9 barrels, or 0.51% of the world's initial fossil energy content.

The Athabasca tars sands cover approximately 30,000 square miles of Northeastern Alberta. Its sands contain a thick liquid that is resistant to flowing. To separate this bitumen from the sand, the sands will first have to be mined. The sands are generally covered by several hundred feet of overburden. After mining, the sands are processed in a so-called syncrude plant, where the produced bitumen, (about 90% of the contents of the sands) is upgraded in a complicated process that involves adding hydrogen to it. The result is a high-grade crude oil. The remaining sand, from a single plant, would make 380,000 tons of waste every day that would have to be disposed of. Large amounts of water are added to the sand during processing. It would be a major environmental problem to dispose the resulting mess.

The tar sand's oil resources in Canada will probably not be fully developed for production before the end of this century. The construction costs of a single plant, producing 125,000 barrels per day, is put at about $800,000. At remote locations, labor force for the construction is scarce. Building only one plant takes about five years, with the work of about 3,000 laborers at its peak.

Recent conservative estimates claim that Canada's tar sand oil reserves are about double that of the Arab oil fields, making this source one of the most important of the world. Another major tar sand deposit of the world was found in the Orinoco valley in Venezuela. Brazil, on the other hand, is rich in oil shales.

The development of both shale oil and tar sand oils will result in an increase of oil prices but also will bring about a decreased dependence by the United States on overseas suppliers. A crash program on shale oil development would put the United States into a better bargaining position with the oil-rich countries and hence it would alleviate the unfavorable balance of trade deficits. Thus our shale oil resources will improve the domestic oil situation both directly and indirectly.

The importance of oil as a basic material for the chemical industry, aside from its primary position as an energy resource, should also be mentioned. About 5% of US oil consumption goes for the manufacturing of plastics, tires, fertilizers, fibers, and food additives. From these standpoints, the burning of oil as an energy resource is considered a waste.

As described in the previous chapter, oil may be manufactured from coal at a cost that may easily make the process economically feasible in the near future. Federally sponsored studies also led to a new process that converts lumber wastes to low sulphur fuel oil. This process, now in the pilot stage, uses high temperature and pressure steam and carbon monoxide. Eighty-four gallons of oil can be produced from each ton of dry lumber waste by using this process. Ultimately, any organic refuse could be used to make oil if the economic situation warrants it.

CHAPTER 6

Natural Gas

Natural gas is closely related to petroleum. It is formed under the same geological conditions. Decomposition of organic matters in the absence of oxygen, with the aid of bacteria, like the "marsh gas" of swamps, results in methane and other hydrocarbons. If collected and retained in the traps of reservoir rocks, natural gas in large quantities occurs together with oil as well as alone.

Chemically, natural gas is mostly methane (CH_4), about 75% to 99% of the gas. Small quantities of other hydrocarbons may also be found in it, along with carbon dioxide, hydrogen, nitrogen, carbon monoxide, hydrogen sulfide, etc.

One distinguishes between oil-associated gas, which is dissolved in oil at the high pressure in the reservoir and is separated from the oil as it is extracted from the well, and nonassociated gas (or dry gas), which is obtained from gas wells.

Natural gas is measured in volume—at standard pressure and temperature. Units of 1000 cubic feet or higher are used. The energy content of one cubic foot of dry natural gas is 1030 BTU.

Gas being a compressible fluid, it is compressed into a relatively small volume in the reservoir. Because its density is much smaller than that of either oil or water, it occupies the highest points of oil reservoirs. Gas in oil reservoirs aids the extraction of oil in a way similar to the air in pressure tanks of suburban well-supplied home water systems. Gas expands as the pressure is released by pumping out the oil, and it squeezes out the oil from the pores and cracks of the rock. In secondary oil recovery methods, gas is often pumped back into the reservoir to increase the oil production.

67

In the early experiences of oil drilling, gas was considered to be a useless product. Hence it was burned off or was allowed to escape into the atmosphere. Even in our days, the various oil-producing nations are "flaring" some 4.5×10^{12} cubic feet of gas each year as a waste product, which equals almost a quarter of the US gas production in 1970. Utilization of natural gas came long after "city gas" (which was made of coal) was introduced in 1816. Only after the first gas pipelines were built, in 1930, was the energy of natural gas utilized. This progress required the development of new technology in highpressure pumping, storing, and transportation, all of which took decades to perfect.

The introduction of natural gas into the energy market was an important development. Gas is a clean burning fuel, not like coal, which produces enormous air pollution. It is an efficient fuel because it is used in its natural state and does not have to be generated from other fuels. It does need pipelines and storage facilities, which are expensive to construct, but the high elasticity of gas offers some advantages. Depleted gas fields, as well as the pipeline network itself serve as storage. If the pressure increases at the pumping station, the gas flows faster and the storage capacity of the network increases as well. A three-foot diameter pipe of one mile length stores one million cubic feet of gas if the pressure is 400 pounds per square inch.

In the early 1970s, 33% of American energy needs are being supplied by natural gas, versus 44% that are supplied by crude oil. Since 1956, United States gas production has almost tripled. Between 1970 and 1985, domestic gas demand will rise from 22×10^{15} to 40×10^{15} BTU per year. Gas consumption in the United States increases by 4% to 5% each year; 150 million residential, commercial, and industrial consumers depend on natural gas. Over 35 million American homes are heated with natural gas. More than half of all American families cook their food on gas. Gas used by electric utilities amounted to 306.942 million cubic feet in 1946. This rose to 3,982,720 million cubic feet or 26.8 cubic miles by 1971. In 1970, about 26.5% of our electric energy production (equivalent to 15.1×10^{15} BTU) was produced from natural gas. Yet the supply of natural gas is not endless (117).

Of the total initial fossil energy content of the world, it is estimated that gas amounts to only 2.94%. This represents about 10^{16} cubic feet, or 67,500 cubic miles. Natural gas may be the first energy source to be of short supply. It is probable that after 1980 the net gas supply will begin to dwindle. By that time, yearly consumption will be 22×10^{12} cubic feet, or 149 cubic miles, in the United States and the "proved reserves" will be perhaps 280×10^{12} cubic feet, equal to 1890 cubic miles or so—enough for 12 to 13 years.

To shed more light on this problem, here are some additional hard facts:

United States reserves and production of natural gas (83), in trillion (10^{12}) cubic feet per year, were as follows:

Year	Reserves	Production	Reserves/Production
1950	184.6	6.9	22.8
1960	262.3	13.0	20.2
1970	264.7	21.8	13.2

The last column shows that the number of years for which our reserves would be enough to supply subsequent years has steadily declined in the past.

Between 1960 and 1970 wellhead gas production has increased from 13×10^{12} cubic feet to 22.3×10^{12} cubic feet. This was made possible by the large backlog in reserves.

The ultimate availability of United States gas is estimated to be 1250×10^{12} to 2318×10^{12} cubic feet or 8450 to 15,700 cubic miles. This certainly will be consumed by the year 2200. Assuming the ultimate availability of discoverable United States gas to be 1857×10^{12} cubic feet of which 679×10^{12} cubic feet was discovered by the year 1971, in the United States we have 1178×10^{12} cubic feet or about 66% of gas that still remains to be found.

Demand for natural gas increases in this country year by year. On the basis of past performance, it is expected that supply will fall short of demand by the following rate (53):

Year	Demand	Supply	Shortage %
1970	21.82	21.82	0
1975	30.27	19.80	33%
1980	34.70	17.47	50%
1985	40.12	14.50	64%

The above figures are in trillion (10^{12}) cubic feet per year. The last column shows that the gas shortage will be increasing as the years go by. This is in spite of the fact that at least 26×10^{12} cubic feet of gas in an oil-associated state and an additional 5 trillion cubic feet of dry gas is waiting to be tapped on the Alaskan North Slope. To get this gas into the US market, a 2600-mile pipeline would have to be built at a cost of $5.7 billion. It would carry 4.5 billion cubic feet of gas per day, perhaps underground in a refrigerated condition so as not to melt the permafrost. This would supply a mere 6% of the United States' needs after 1980 but it would last only a score of years.

One of the causes of the gas shortage is that gas prices at the wells are federally regulated and are kept at an artificially low level. In early 1970 the price level of natural gas was set at 17.1 cents per thousand cubic feet. This will have to rise to about 40 cents per thousand cubic feet in order to make production and exploration financially feasible. The adverse economic circumstances show up in the number of wells drilled over the years. In 1956 a total of 58,000 oil and gas wells were drilled. This figure has fallen to less than 30,000 by 1970. Gas drilling rates have fallen by one-third between 1960 and 1970. Gas finding rates were falling also, since most "easy" reservoirs are already tapped. In the 1950s 350,000 cubic feet of gas were found for every foot of gas well drilled; in the 1960s only 280,000 cubic feet were found for every foot of drilled wells.

Europe's future supply of natural gas seems somewhat more assured. Although local sources are relatively scarce, the new North Sea find, discovered in 1959 off Holland, will add about 50 trillion (10^{12}) cubic feet to Europe's reserves. In addition to this, Siberia's new Nadym gas field, which contains 212 trillion cubic feet of gas (equivalent to about two-thirds of present United States' reserves), will be sent to Western Europe by 1978 after a new 600-mile gas line is completed. American companies are considering a $7 billion 2000-mile pipeline to the Port of Murmansk for shipment of gas in a liquefied form to the United States' East Coast.

To solve the emerging gas supply problem in the United States, first a reorientation of gas utilization seems to be in order. United States gas requirements are shared presently in the following manner:

residential use	23.3%
commercial use	9.4%
industrial use	27.9%
electric generation	39.4%

Gas-fueled generating stations became very popular in the recent past. It is the present policy trend to conserve gas for superior uses in which residential use has first priority, and to restrict the use of gas as boiler fuel in electricity production where other kinds of fuels are available.

Supposedly there are about 3 trillion (10^{12}) cubic feet of natural gas locked beneath the Rocky Mountains. The United States Atomic Energy Commission is sponsoring a series of tests to see if releasing this gas by nuclear explosions is economically feasible. Initial results of this "Plow-share" study increased the gas yield of test wells by a factor of 20. For stimulating the production of a number of wells in Colorado's Rio Blanco valley, several nuclear bombs of 30 kilotons range were simultaneously exploded in the spring of 1973. However, to release all gas under the Rockies, some experts claim that about 370 nuclear blasts would have to be made annually until the end of the century—a blast each day for over 25 years, requiring perhaps more energy than the energy content of the gas gained. Environmental objections to this project range from citing dangers of ground water contamination to endangering the oil shales located above the area in question. Experience has shown these objections to be largely unfounded (58). Figure 6.1 shows some major basins within the Rocky Mountains from where the thicknesses of potential gas reservoirs warrant the development of stimulated gas fields using the methods developed by the Plowshare program, but the most recent experiments showed discouraging results.

Transporting natural gas in a liquefied form in special tankers is already practiced in the form of Algerian gas sent to the United States. Liquefied gas is made by cooling the gas to a temperature of 260°F below zero. In this liquefied state gas takes up only 1/600th of its original volume. Of course this has both storage and transportation advantages, making it possible to supplement the United States supply of natural gas from overseas sources. Transporting Siberian, Libyan, and other gases is being considered. Liquefied natural gas, however, needs expensive special facilities such as liquefaction plants, special tankers, regasification facilities, and storage. The cost will probably keep the refrigerated gas supply restricted to a relatively small, high-priced market. Critics consider these special refrigerated tankers to be floating bombs. In case of accident the released gas could create an enormous fire-cloud that could destroy large coastal areas. Newspaper accounts of the October 20, 1944 explosion of America's first liquefied gas tanks in Cleveland, Ohio, are grim reminders of what could happen. More than 25 million horsepowers were released by the cracking of a tank when its metal was made brittle by the supercooled gas. With flames towering 2800 feet and heat waves of about 3000°F temperature more than a square mile of Cleveland's East Side was devastated, killing 131 people and injuring hundreds of others.

Of oil-associated gases, the lightest produced in oil refineries are methane and ethane. These normally serve as fuels in the refineries and may be burnt as waste products if not saleable. Propane and butane, which are denser, are first scrubbed and then liquefied under pressure and

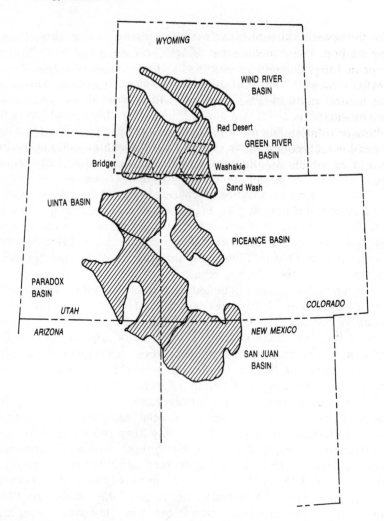

Fig. 6.1 Map of Rocky Mountains gas fields considered for "Plowshare" development. (After Ref. 2.)

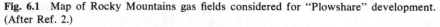

stored in tanks before placed in steel bottles for industrial and domestic use.

Substitutes of natural gas will play an increasingly large part in the future. Crude oil fuel preparation complexes will prove to be the most extensive means for supplying these gas substitutes. By using crude oil as a base, these fuel preparation complexes can supply the naphtha fraction of oils for the several substitute natural gas processes using the catalytic reforming method. These include the British Gas Council's CRG process, the European Lurgi Gasynthan process, and the Japan Gasoline Company's MRG process. The gas produced by these methods is a true substitute natural gas with a heating value of 1000 BTU per cubic foot after enrichment.

The ultimate solution for the coming gas shortage rests on advanced coal gasification techniques due to the fact that coal is the only fossil fuel in the world of which adequate resources will be available for many centuries.

CHAPTER 7

The Heat Energy
of the Underground

Mankind has been well aware of the fact for a very long time that the inside of the Earth is a very hot place indeed. The eruptions of some volcanoes have been chronicled for over 2000 years in Japan and Italy. The geologic record is replete with evidence of many eruptions of over 300 volcanoes around the world.

Sites where hot waters come from the Earth are in the tens of thousands around the world. Natural springs, "hot springs," and "warm springs" in the continental United States are known to exceed 1000. Artesian wells bring up waters near the boiling point at many places around the Earth. In deep oil wells along the Gulf Coast temperatures exceeding 273°F have been registered.

The increase of the Earth's temperature with depth is documented in deep mines and long tunnels. The highest rock temperature measured in the Alpine Simplon tunnel exceeds 55° Celsius. It is known that at a certain depth below the Earth's surface the temperature and pressure reach a level at which the rocks are in a liquid, molten state. This molten rock is called magma. If the magma breaks into the solidified outer crust of the Earth, it crystallizes and forms so-called igneous—fire made—rocks.

The solid crust of the Earth is broken up into rigid continental shields forming the central flat parts of continents. Caused by extraterrestrial events, these shields tend to shift with respect to each other. In this process the regions around the edges of the continental shields are weakened by faults and bends in the Earth's layers. Through these weakened regions the hot molten magma of the Earth's depths may intrude toward the surface. These "hot spots" are detected not only by their outward signs as volcanoes or natural hot springs or geysers like those at Yellowstone Park, their presence is also indicated by the relatively rapid increase of temperature with depth when wells for oil or water are drilled. The rate of increase of the Earth's temperature with depth is called the thermal gradient. It is expressed in degrees Celsius per thousand meters. The normal thermal gradient of the Earth ranges between 8°–50° Celsius per thousand meters, averaging at 25°C/km. The outward flow of heat energy from the Earth's depths exits at a world wide average rate of 1.5 microcalories* per second on each square centimeter of the Earth's surface. This is a very small amount when compared to the sun's incoming energy, but it is a potentially very important energy source. There are regions of intense geothermal manifestations in the areas around continental shields like America's western and central coastal areas, Japan, the Chinese coastal regions, the Pacific Islands, New Zealand, the areas of the Mediterranean coast, Central Europe, Africa's eastern coastal regions, the Mideast, Iran, southeast Asia, and other sites shown in Fig. 7.1. Geothermal gradients at these regions may range from 100° to 300° Celsius per thousand meters. This indicates that there are immense areas on the Earth's surface where underground temperatures are concentrated such that they reach five to ten times their normal average value. These areas are called thermal reservoirs. The heat energy of these thermal reservoirs can be converted to useful work by existing technology in many cases (70).

Before we proceed, we may ask, "Where does this heat come from?" When the Earth was formed as a star it was at an extremely hot temperature. Since that time it should have cooled down, since it is losing about 20 million kilogram calories through each square mile of its surface every year. The heat loss is, fortunately, being replenished by the heat generated through the fission of nuclear materials in the magma. The small amounts of fissionable uranium,* thorium, and other such elements in the rocks maintain the flow of heat from the inside of the Earth at a constant rate. Based on the fissionable content of an average granite (a common igneous rock), it has been estimated that an 18-kilometer thick granite shell around the Earth would generate the amount of heat radiated and lost under natural conditions, some 32×10^{12} watts (121). The development of geothermal energy as an important national energy resource

---

*One microcalorie equals one-millionth of a calorie.
*More on nuclear fission in Chapter 11.

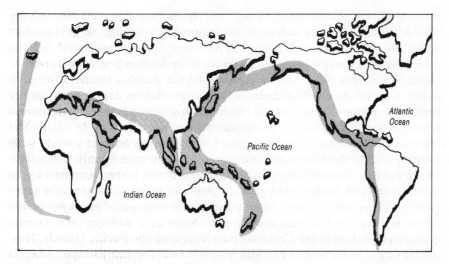

Fig. 7.1 Map of regions of intense geothermal activity around the world. (After Ref. 70, Copyright © 1973 by Stanford University Press. All rights reserved.)

depends entirely on the extent to which geothermal heat in its more abundant forms—the thermal reservoirs—can be located and extracted economically. Basically, the energy is there in immense quantities (92). Theoretically, this energy can be tapped from any point on the Earth by simply drilling sufficiently deep bore holes and allowing a heat transfer fluid to circulate between the hot rock mass and the surface. Suitable heat exchangers could transfer the heat energy to a turbine or other system that can convert the heat into electrical energy. Of course there are practical limits to the depth of drilling, so one may assume that only an upper layer of, say, 10 kilometers may be exploited. Assuming a thermal gradient of only 20°C per one kilometer, scientists claim that a total energy of $6,000,000,000,000,000,000,000,000 = 6 \times 10^{21}$ kilocalories is stored under the United States. This equals 70×10^{12} megawatt-hours of energy. Over a year, this would provide 612×10^9 million megawatts of power. Comparing this to the 0.34 million megawatts used in the United States in 1970, one realizes the magnitude of the Earth's thermal energy resources (70).

Most of this staggering amount, of course, will remain locked in the Earth because of the excessive cost necessary to retrieve it. Fortunately, we do not need more than an extremely small fraction of this energy reserve. Recent studies indicate that with the technological knowledge existing or within our reach by a relatively modest research program we would be able to produce one billion megawatts of energy in less than 15 years and 3.1 billion by the year 2000.

Geothermal energy is known to be used for heating and for residential, commercial, industrial, or agricultural drying purposes at a number of places around the world. Heating with geothermal waters, by transferring their heat through a heat exchanger to fresh water, was applied at several places around the world. In Iceland and Hungary, these thermal waters are used directly for heating. Cost of such heating has repeatedly been proven to be much cheaper than heating by any other known fuel. Because heat dissipates rapidly, hot water or steam from the depths of the Earth cannot be transported very far but has to be utilized at the point where it is produced. The farthest distance in Iceland to which hot water from wells is piped for home heating is less than 15 miles. For electric energy generation, piping should be less than one or two miles.

The heat content of the Earth's depths can be extracted in several forms, depending on the underground geologic conditions. The most developed form is the one that is easiest to tap. This is the utilization of dry underground steam for power generation. Underground steam exists at a few places around the world, most notable of these are the fields of Larderello in Italy, The Geysers in California, Matsukawa in Japan. All of these exist in relatively recent volcanic regions, where high thermal gradients exist, the source rocks have high permeability due to cavernous limestone, porous sandstones or other cracked formations, underground water supplies are large, and superheated steam is allowed to develop. The world's largest geothermal steam reservoir known today is the one at The Geysers in California. Small-scale local utilization of this site for electric power production was begun in 1920. Large-scale utilization began in 1960. By 1974, the capacity of this plant was 522 megawatts. An orderly development of the generating capacity is planned. According to plans, the capacity will be raised to 1180 megawatts by 1980. The maximum potential production was estimated to be as high as 4800 megawatts.

The first geothermal steam-operated electric power station was installed in 1913 at Larderello, Italy (70). From the initial 250-kilowatt generator, this plant has grown to a generating complex of 13 plants, producing 390 megawatts of electricity. These reservoirs are found at about 1000 feet below the surface. The temperature of the steam is 240°C. A similar plant in New Zealand produces 8% of the nation's electricity from the geysers of Wairakei. The steam is taken from 60 bore holes ranging in depth to 4000 feet.

The conversion of geothermal steam into electric energy is simple. It requires the gathering of the steam in insulated pipes from the wells into the power station and allowing the steam through a largely conventional steam turbine. The conversion efficiency of geothermal steam-type power stations is about 14%. Steam from geothermal sources does contain some other gases. From an environmental pollution standpoint, H_2S stands out. Still, when compared to the hydrogen-sulfide production of coal-fired plants, potential geothermally induced air pollution is negligible.

As underground steam systems are easy to utilize with existing technology, they are the most commonly tapped forms of geothermal energy. About 73% of the world's total geothermal energy production comes from steam-type reservoirs. But these types of systems are very rare, perhaps less than 5% of all systems with temperatures about 200°C.

Much more prevalent are the hot-water reservoirs. These are characterized by permeable rock layers through which surface waters enter the lower layers of the Earth. There they come in contact with a mass of hot magma. Through the heating effect of this base rock, the temperature of the water may rise to as high as 300° to 380°C. At these temperatures, water is able to dissolve a large number of minerals such as silica, chlorine, boron, potassium, sodium, lithium, and many others. In the Imperial Valley of California sodium-chloride brines of high temperature (250° to 370°C) are relatively common. Commercial utilization of these chemicals is feasible in many cases. There are some major problems in the utilization of these geothermal fluids—namely, that the high corrosive quality of the fluid requires expensive corrosion-resistant pipes and equipment; also, the effluent may create intolerable water pollution problems if allowed to enter surface waters.

Utilization of hot-water geothermal reservoirs for electric energy generation is possible in the form of a closed loop heat exchanger system. If the underground hot water is extracted through bore holes and let through a heat exchanger, it will transfer some of its heat to clean water. Steam produced in the heat exchanger may then drive a conventional steam turbine. After the steam passes through the turbine, it may be condensed, cooled, and readmitted into the heat exchanger. The geothermal brine may be pumped back into the underground reservoir after it passes through the heat exchanger and its valuable mineral content is partly removed. By this method, the polluting effect of the underground brine is eliminated from the surface. At some locations around the world, where pollution does not appear to be a problem, hot waters from the Earth drive electric generators directly. In most cases part ($\sim 25\%$) of the water turns into steam with the reduction of the pressure. The steam is then allowed to drive steam turbines. Still, corrosion is a problem and the efficiency of the conversion is much worse than that of dry steam. Where the need exists, it is possible to desalinate some of the hot brine and produce fresh water for municipal purposes (70).

Where the underground waters are only moderately hot (80° to 150°C), steam will not form and hence conventional steam turbines are inapplicable. Such lower temperature geothermal fluid reservoirs comprise the overwhelming majority of fluid-based geothermal fields. Huge areas of the world are known to exist where such moderately hot waters were found in relatively shallow wells in great abundance. Almost half of Hungary, large areas of the central and western parts of the Soviet Union, and at least a dozen other countries have such moderately hot waters available. Production of electric energy from such waters is possible by a new technique using a "vapor cycle turbine." This method uses a heat exchanger to transfer the heat contained in the well water to a low boiling point fluid, like freon* or isobutane. At the temperature obtained from the water, these fluids flash into vapor. This vapor is allowed to pass through a specially designed turbine that is much like a steam turbine. After leaving the turbine, the fluid is recondensed and allowed to reenter the heat exchanger as shown in Fig. 7.2. In this manner the vapor cycle is a closed loop system (70). Design of such vapor cycle turbines is a much simpler task than the design of steam turbines used today. The geothermal waters can be pumped back into the deep layers, eliminating both the potential air pollution by the gases contained in the underground water and water pollution through the chemicals held in it. Due to its relatively high efficiency, the vapor cycle turbine system offers perhaps the greatest promise for utilization of the energy held in the abundant supply of hot underground waters. Further research for improved equipment and materials would well be worth the effort.

Most heat reservoirs under the surface are devoid of water. There are areas covering hundreds of square miles that contain immense volumes of hot rock at temperatures between 300°C and 600°C located in relatively shallow depths of less than two miles. In fact, in most areas of the United States, perhaps only with the exception of the north-central states, the average rock temperature at depths reachable with today's drilling technology (25,000 feet) is upward of 150°C. If techniques were available to extract the recoverable energy from dry, hot rocks the energy needs of the United States would be assured for about 50,000 and up to 500,000 years if used at the present rate of consumption. The technology capable

*Registered trade name of Du Pont Co.

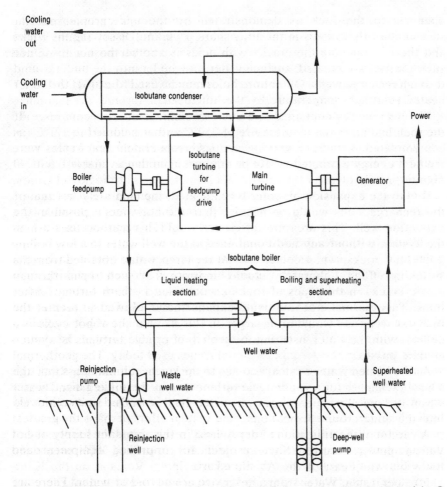

of capturing this immense source of energy is not far from our reach.
Some very feasible proposals were advanced recently. These are based
on several new techniques that were successfully demonstrated in the
past few years.

Extracting heat energy from dry, hot rocks would be possible by
drilling holes into the hot rock mass and creating cracks and cavities by
any one of the methods used in conjunction with oil and gas well
stimulation. Blasting conventional or nuclear explosives would generate

openings in the rock as demonstrated by the underground nuclear stimulation of gas wells in the Plowshare program. Dissolving the rocks and thereby opening the cracks with acids is another possibility. Once such cavities are created, surface water may be let into the underground through recharge wells. Other bore holes may be used to extract the water heated to a high temperature by the hot rocks.

The heat energy contained in a one cubic mile is impressively large. If the rock had an initial temperature of 350°C and was cooled to 177°C (as contemplated in studies involving cavities created by nuclear explosives), the heat energy extracted would be equivalent to that available from 300 million barrels of oil.

Due to the expansion of water when heated, the cold water column of the recharge wells would be heavier than the hot water column in the extraction well. This pressure difference would drive the water through the system without any additional pumping.

The hot rocks, when cooled by the recharge water, would undergo a shrinking effect. Laboratory studies indicate that such shrinking may create cracks in the types of rocks encountered in the hot rock formations. This thermal cracking could expand the cracked zone around the recharge wells and would bring larger and larger volumes of hot rocks into contact with the water, assuring the growth of productivity of the system as time passes.

As the heated water or steam comes to the surface, it may pass through a heat exchanger for a vapor cycle turbine or may be utilized directly in a steam turbine. The spent steam or water may be cooled and reinjected into the underground formations.

A variant of the utilization of dry thermal reservoirs is the harnessing of volcano power. A current Soviet project calls for drilling 3000-meter deep holes into the heart of the Avachinskaya Spoka Volcano on the Kamchatka peninsula. Water from a nearby river will be pumped into some of the holes, while other holes will serve as steam outlets. The steam will be transferred to turbines, which will produce electrical power.

Scientists have calculated that if all the world's electric demand would be met in the future by geothermal energy alone, it would take 41 million years for man to reduce the temperature of the Earth by one degree Fahrenheit. There is much more than enough heat available inside the Earth to generate all the power we need even if we maintain the present rate of growth in energy utilization. And we might need only a fraction of the cost of our past space program to develop the necessary technological base to make this ideal method of energy utilization a reality.

CHAPTER 8

Power from Sunshine

Most of the energy we can account for on the Earth originates from the sun (106). Perhaps the only exception to this is nuclear energy, whether generated in a nuclear power plant or in the interior of the Earth as manifested by geothermal power. The energy of the sun radiates toward the Earth in the form of sunshine—an electromagnetic ray (132). One day of sunlight on the surface of Lake Erie is about equivalent to one year's worth of energy used by the whole United States (61).

Light consists of bundles of energy called photons. The energy content of a photon is proportional to the frequency of the light—the higher the frequency (shorter the wavelength), the higher is the energy content. Blue light at a wavelength of 450-billionth of a meter has an energy content twice that of infrared light, which has a wavelength of 900-billionth of a meter. Ultraviolet light with a wavelength of 225-billionth of a meter has four times the energy of infrared light.

The solar energy reaching the Earth's atmosphere amounts to 1.395 kilowatts per square meter. Taking the surface of the Earth on its diametric plane to be 1.275×10^{14} square meters, the total power reaching the outer layers of the atmosphere is therefore 1.73×10^{14} kilowatts.

The Earth's atmosphere filters sunlight by absorbing most of the high-energy ultraviolet light and some of the infrared light. About 30% of the incident solar energy is directly reflected and scattered back into space as light, a short wavelength radiation. Another 47% is absorbed by the atmosphere, the land, and ocean surfaces and converted to the form of long wave radiation—heat. An additional 23% is utilized as the energy source of the hydrologic cycle, driving the processes of evaporation, precipitation, winds, ocean currents, and waves; in this form it is eventually dissipated by friction-generated heat (62). This is the energy source of hydraulic power stations, wind energy, heliothermal and hydrothermal energy schemes. A very small additional fraction, mainly composed of blue light and red light, supplies the energy needed in photosynthesis in the process of plant growth. Photosynthesis fixes the atmospheric carbon dioxide in the vegetation and produces hydrocarbons. The minute fraction of these plants over the whole history of the Earth decomposed in the absence of oxygen on the bottom of water bodies. These formed our coal, oil, and natural gas deposits.

The amount of solar energy entering our environment—insolation, as it is called—along with the amounts of it that control the weather or turn directly into heat, is in precarious balance. Burning of our fossil fuels and generating heat by the necessary cooling of nuclear reactors result in excess heat that enters our waters and atmosphere. In our days the total energy consumption of the United States constitutes about 0.2% of the solar energy falling on its land area. But with the present rate of growth, it may be only 50 years until the waste heat generation of the United States is sufficient to cause an increase in the average environmental temperatures by more than 2.5°F. In a few places, such as New York City, the man-made energy contribution is already greater than the normal influx of solar energy. On a continental scale, atmospheric effects will reduce the heating effects (as it happens today in New York) but will not eliminate them. With continuation of the current growth of thermal pollution, the global heating effect will reach significant levels at which it will have the potential to influence weather. This, in turn, could endanger our very life by wreaking havoc on the natural balance of agriculture.

Proponents of solar energy utilization point out that while combustion of fossil fuels and freeing of nuclear energy both represent such heat energy contributions, solar energy can be used without any thermal pollution. It keeps coming in, whether we want to use it or not.

The incoming sunrays that keep our environment and atmospheric conditions in such a precarious balance do, in a large part, diffuse heat into our atmosphere. Atmospheric pollutants, both particulates and gases, change the transmissibility or, in other words, the heating effect of the air. Solar energy measurements in downtown Boston show almost 20% reduction of insolation as compared to the outskirts of the city. Increasing global atmospheric pollution may endanger life, as we know it, by reducing the incoming solar energy and altering the energy balance of our Earth.

In view of the fact that our fossil fuel reserves are final and at the present rate of use they will be exhausted within a relatively short time, a number of energy experts recommend an aggressive development program for the direct utilization of solar energy (80, 115).

Indirectly, solar energy is already being utilized. Agriculture as well as fossil fuel utilization make use of the sun's energy through photosynthesis of the past and present. Water power development from the waterwheels of the old days to Grand Coulee and other dams (or windmills) converts solar energy through photothermal processes, since it takes the atmospheric beating effect of the sun to evaporate the oceans' water, create air currents, and make rains fall over the land. But direct conversion of solar energy to produce electricity has scarcely been tapped (86).

The most favorable sites for the direct utilization of solar energy or for large-scale development of solar-electric power are desert areas not more than 35° north and south of the Equator. Such areas are to be found in the southwestern United States, in the Sahara, the Kalahari desert, the Arabian peninsula, central Australia, the Atacama desert of northern Chile, and others. These areas receive 3000 to 4000 hours of sunshine each year, delivering up to 650 calories per square centimeter per day. Even in the winter, 300 calories of insolation averaged over 24 hours supply about 145 watts on each square meter (15). Figure 8.1 shows winter and summer insolation over the United States.

Unfortunately the largest amounts of solar energy are found on the least developed regions of the Earth (74). From the standpoint of developing these regions, solar energy utilization is already feasible. Small-scale power schemes—like solar-powered pumps to raise water from wells, distillers to provide fresh water by desalinization of sea water (123), heating water for household use, house heating, air-conditioning, refrigeration, cooking, and small-scale production of electricity—have been demonstrated to be fully feasible (131).

The utilization of solar energy is not a recent development (19). Archimedes burned Roman ships besieging Syracuse in 212 BC with the sun's rays reflected and concentrated by the shields of hundreds of soldiers. In the 17th century E. W. Tschirnhaus used a parabolic copper mirror of one meter radius to kiln-dry porcelain. In 1875 a 20-square-meter mirror drove a steam engine generating one horsepower. A solar energy desalinizator built in 1872 worked in Chile for several decades. Industrial solar furnaces built in France, and recently in New Mexico, concentrate sunrays by large mirrors, producing temperatures up to 5000°C (enough to melt even the most heat-resistant materials).

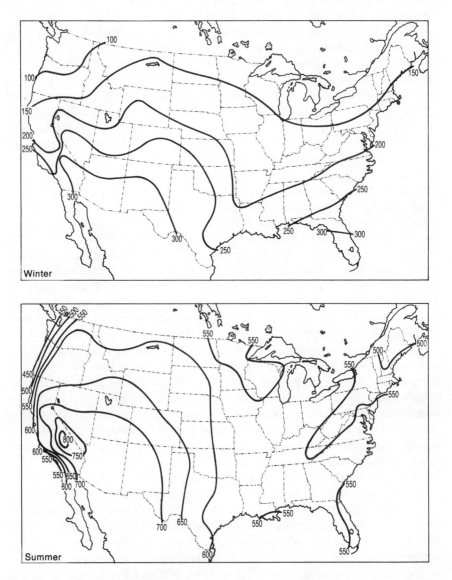

Fig. 8.1   Maps of winter and summer insolation over the continental United States. (After Ref. 15, courtesy of Pergamon Press.)

Sosses, a scientist in Sweden (1740–1799), recognized the so-called "hothouse effect." A glass panel covering the top of a black box allows the sun's rays to enter and increases the temperature in the box by retaining the reflected heat. Air in a well-insulated box may be warmed to 80°C. Using seven layers of ordinary glass or double glazing with selective surfaces increases the efficiency; it can generate 200°C temperature in the box. Such systems are used in many parts of the world to provide hot water for homes. In the southern parts of the USSR water heaters with a surface area of five square meters provide the hot water requirements of whole families.

In the field of electric energy production in developed countries much more substantial schemes are needed in order to replace fossil fuels on an industrial scale (128). Utilization of solar power to date was limited to developing photovoltaic energy converters, solar cells, to be used in spacecraft. A photograph of America's Skylab, showing its solar panels, is depicted in Fig. 8.2. To translate this space technology to large-scale solar electric power generation is not a simple matter. There are major technological difficulties to be overcome before the electric energy used in the United States can be of solar origin (61).

Fig. 8.2 Skylab with its extended solar panels. (Courtesy of the National Aeronautic and Space Administration.)

Presently, direct conversion of sunlight to electricity is done by silicon solar cells. At normal solar radiation levels, of say 90 milliwatts per square centimeter, such cells provide about 12 milliwatts of electric energy per cm^2. The cost of high purity silicon crystals used in spacecraft today is very high. To produce only one kilowatt of electricity, over $3500 must be invested in silicon alone. It would therefore be necessary to develop new techniques for photovoltaic converters, solar cells, before their energy will be competitive with conventional energy production methods.

Experiments show that by using lenses and mirrors to concentrate sunlight on solar cells, their output can be increased significantly. Power output of conventional solar cells can be increased to 1.5 watts per square centimeter with the use of such solar concentrators. But the resulting high temperature at the cell requires a cooling system to maintain an acceptable operation temperature. The continuous mechanical adjustment of the concentrators to follow the position of the sun adds additional costs to the total system.

On the Earth, sunshine is an intermittent source of energy. Daily variations, the expectability of a number of consecutive cloudy days, and significant seasonal variations of insolation intensity make the problem of storing energy for "a rainy day" an acute one. Conventional batteries are thoroughly insufficient for this purpose. One promising new technology that has been identified as applicable to energy storage is the use of high-performance rechargeable batteries. These would serve a purpose similar to that of pumped storage in hydraulic power generation. One example of a promising high performance battery system is one employing lithium and sulfur, utilizing a molten salt electrolyte. This system has very high energy storage capabilities, about five to ten times that of the conventional lead-acid storage battery used in automobiles. Similar high-performance systems, employing sodium rather than lithium as anode, are also under development both in the United States and abroad. Studies on the lithium-sulfur system, for example, are still in the experimental stage in major research laboratories. Only a few single cells are operating, delivering a few watts of current. Although a significant research, development, and engineering program will be required to develop a megawatt-sized battery, there appear to be no insurmountable technical barriers to achieving this goal. To show the complexity of even such a small segment of the whole developmental problem related to solar energy, a list of some important unsolved technological problems follows:

search for lightweight, inexpensive corrosion resistant materials,
development of high-temperature seals and feedthroughs
improvement of cathode structures for higher electrical capacity density and longer life,
development of an electrolyte with optimum wetting property and low
solubility for sulfurous materials, and
development of an optimum battery design.

Once such problems are solved, a battery of one kilowatt capacity, envisioned to be about the size of a seven-foot cube, could be completed in a few years. Commercial demonstration of energy storage systems utilizing such batteries may come by about 1980 (2,3).

If electricity storage can be done on a large scale, the difficulties resulting from the intermittent nature of terrestrial solar energy production can be overcome (73).

Various proposals were made by advocates of solar energy utilization to develop demonstration-size solar power stations (13). One proposal calls for a one square mile area around Phoenix, Arizona, to be covered by lower cost, newly designed silicon solar cells, connected in a weather-protected array. At 10% conversion efficiency, this would produce 1.6×10^6 kilowatt-hours of electricity per day. The produced direct current would have to be converted to alternating current and could be connected to the existing generating system. The proposed scheme would require a large amount of silicon, and a considerable development in silicon and cell manufacturing technology. With this type of generating system, a composite area of 100,000 square miles of solar cell fields would be required to supply all the electric energy projected to be needed by 2040—a mere 65 years from now.

Another proposal for large-scale solar power generation calls for a reflector system of a square mile area, projecting solar radiation onto a solar furnace and boiler at the top of a 1500-foot tower. Heat from the boiler at a temperature of 2000° Kelvin would be converted into electric power by a magnetohydrodynamic (MHD) method* yet to be developed. It is based on passing hot gas, seeded with conductive particles, through a magnetic field. The concept was discovered back in Faraday's time but was not studied until the 1950s. An energy storage system based on

*More on the MHD in Chapter 13.

hydrolysis of water to produce hydrogen fuel is proposed. Overall conversion efficiency is said to be 20%.

Yet another proposal for terrestrial solar energy conversion calls for the utilization of the so-called "hothouse effect" in thermal collector pipes. This involves the use of films that absorb most of the incident solar radiation but reemit little infrared radiation. The proposal calls for tubes running in an east-west direction conducting molten sodium and potassium through a circulating system. The tubes are enclosed in glass pipes from which air is evacuated to protect a selective radiation film on the tubes and to reduce heat losses from the tubes. Either by reflector-concentrators or by Fresnel lenses, sunlight would be concentrated on the tubes, generating an operating temperature on the tubes of 540°C. The hot liquid would be collected into heat exchangers. The heat could either be used to operate a conventional steam-electric power plant or could be stored at a constant 1000°F temperature in an insulated chamber filled with a mixture of sodium-potassium chlorides of enough heat capacity for at least one day's collected solar energy. This proposal would have an overall conversion efficiency of 30%, claim its proposers. A 100-megawatt power station with such a scheme would require a collector field little over a square mile in area, and would require a 300,000-gallon thermal storage tank. Proposers recommend building it at Yuma, Arizona. Although the initial outlay of money would be many times the cost of a nuclear power station of equal size, it would, according to its inventors, be able to supply energy at a cost of only one-half cent per kilowatt-hour—a remarkable price at today's energy costs. Demonstration of this system would be a worthwhile engineering development. By cheap production of the collector components in large quantities with assured long life and relatively maintenance-free operation, the proposal is within the range of today's engineering feasibility (96).

Much of the energy of solar radiation is lost after the light traverses the Earth's atmosphere. The idea occurred to some scientists that if a solar energy collector would be placed into orbit outside the Earth's atmosphere, it would be able to collect far more energy than those on Earth, which are subjected to the rotation of the Earth and to inclement weather. Furthermore, if such a satellite were placed in a synchronous orbit of appropriate position, it would be capable of receiving sunlight 24 hours a day. As a result, a satellite system would be capable of collecting six to ten times the energy of a terrestrial collector (3).

This proposal recommends two identical satellite collectors, each five miles square, in synchronous orbits 22,300 miles above the Equator, in a position fixed with respect to a receiving station on the Earth, and arranged in such a manner that each can carry a full load while the other is on the dark part of the Earth. These would consist of light-weight, newly developed solar cell panels; operating at an efficiency of 18% they would collect 85×10^6 kilowatts of radiating solar energy. Each of these solar panels would develop 15×10^6 kilowatts of electric power. Through a superconductive cable of two miles in length, they would transmit this power to a control station where the power would be converted into microwave energy for beaming the power to the Earth without incurring losses across the atmosphere. On Earth a receiving station equipped with a receiving antenna of six miles square, would collect the energy and this would be transformed into 10×10^6 kilowatts of electric power, enough to supply the city of New York.

Although the idea captures the imagination in a Jules Verne-like manner, with anywhere near today's technology its construction borders the impossible. It is conceded that if the necessary breakthroughs in technology can be made, success of the concept could be a great value to the power industry.

Feasibility of this concept is dependent on the success of several major developmental features:

solar conversion at 80% efficiency,

cryogenic transmission of high voltage DC from the array to the microwave generator at very high efficiency,

microwave transmission to earth with high accuracy and efficiency, and

conversion of microwave energy to electric power at high efficiency.

After more than a decade and roughly $200 million of development, the average efficiency for silicon solar cells has been increased by less than 1% overall. There is promise, however, that the present 10.5% efficiency can be increased to 14% in production. Unfortunately, the solar array efficiencies degrade in space to about 7.5% when perpendicular to the rays of the sun. Current silicon solar cells produce about 10 watts per pound. New techniques have shown that 22 watts per pound is achievable; the current development goal (conceptual only) is 50 watts per pound (3).

Establishment of a large solar energy power station in a synchronous orbit would require new space booster capabilities. Present equipment, although highly sophisticated, cannot take heavy loads into a synchronous orbit. The power station would have to be assembled in pieces requiring thousands of flights. Based on existing booster capabilities and launch from Cape Canaveral, the unit charge to place equipment in a synchronous orbit is about $4000 per pound. It should be noted that this charge is probably a nominal value.

Based on the best silicon solar cell technology available today (a power-to-weight ratio of 22 watts per pound), the solar cell array alone in the 25-million-kilowatt system would weigh about one billion pounds.

Assuming that the other components in the system would add 25% to the weight, at $4000 per pound in orbit, this would result in a cost of $200,000 per kilowatt to merely place the hardware in orbit for one satellite. This does not include the many thousands of flights needed to erect and man the supporting space station nor the cost and manning of the ground station. With current rocket equipment, this could require a Titan flight every hour around the clock for nearly 100 years. To handle rocket traffic of this magnitude, a launching complex many times larger than Cape Canaveral would have to be built and its cost charged to this power station. The indirect capital costs of $200,000 per kilowatt would account for perhaps one-third of the total cost; interest and escalation during construction would add another two-thirds. In order that there be no interruption of power during the dark portion of the satellite orbit, a second station is proposed. The $200,000 per kilowatt would be increased to the order of $1.5 million per kilowatt. Cost of the ground receiving antenna could equal the cost of the solar collector.

Since the time that man can spend in space has obvious physiological and psychological limits, crews would have to be replaced every few weeks. Operating and maintenance costs for the space power station could amount to a significantly large cost and would be handicapped by the large cost of shuttling crews back and forth to this station at $4000 per pound. Round-trip cost for one 200-pound man would be more than $1 million.

To mitigate these high cost problems, it was proposed that solar cell technology and the balance of the system would operate at near perfect efficiency, and that the power-to-weight ratio could be improved a thousand times or so. This should be coupled to a significant reduction in cost to position and maintain the equipment in synchronous orbits. The cost of development to achieve all these goals could be staggering, with no assurance of success. Clearly, this space power system is far beyond our present or reasonably projected scientific and engineering capabilities.

In addition to this conclusion these are the uncertainties related to the unusual hazards to the public, which this system may represent. The

microwave beam intensity of one watt per square centimeter, as proposed, would be adequate to damage human tissue. This is roughly equivalent to the burn one would receive while holding a lighted 100-watt light
bulb. (It may also be noted that a 1970 FDA standard limits emissions
from microwave ovens to less than one milliwatt per square centimeter.)
A misalignment of one minute of arc in the microwave transmitting
antenna could create a major public hazard, resulting from the beam
missing the receiving antenna on earth by two miles.

It should be realized that, from a thermal pollution standpoint, satellite
collectors do add to the energy balance of the Earth after the energy is
used. It is estimated that with the necessarily heavy governmental support
such a power scheme cannot become a reality until after the year 2000, by
which time our energy needs would have approached a level at which
newer types of energy supply systems must be on line. It would require
250 satellite stations of the type described above to supply the 2500
gigawatt electricity needed by the United States by the year 2000. In all,
this is an impressive but highly impossible proposal.

Unless there are some unexpected breakthroughs resulting in much
more effective photovoltaic devices and much more efficient electric
storage methods on a large scale, late runner solar power will not be able
to catch up with nuclear power, which has been supported over the past
generation by hundreds of millions of dollars every year. In fiscal year
1972, solar energy research received $15 million, while nuclear research
got $560 million.

Without a massive infusion of government research support, solar
energy will be relegated to its present infantile state—being used in
camping cookers, pumps for Bedouins, water desalinizators for
Polynesian natives—as a status symbol of underdevelopment (100). Due
to the 1973 oil scare, the US government allocated a growing share of
energy research for solar energy. The most promising fields for solar
energy utilization are home heating and hot water production. It is
conceivable that within a few decades large amounts of our centralized
energy production will be saved by building solar heated homes, particularly in the southern parts of the United States. Even in northern climates,
solar energy would need only 15% to 20% supplement by nonsolar
standby heaters.

Indirect utilization of solar energy by using commercial forests as
"energy plantations" was also proposed. Forests supplied the major share
of our energy needs until about a hundred years ago. Steam power of the
past was largely based on wood-firing boilers; home heating and cooking

were done by wood. It is estimated that a 400-square-mile forest, efficiently planted, fertilized, and harvested, could supply about 400 megawatts of electric energy a year. There are commercial forests covering about 23% of the land in the United States. Less than one-fifth of our available forests could provide the electric energy needs of America if operated as energy plantations.

There are two ways to convert vegetation into useful energy. One is to construct wood-fired power plants at convenient locations. The burning of wood is nonpolluting. There is less than 0.1% of sulfur in wood. The amount of carbon dioxide generated in the plants equals the amount used by the trees while building their vegetable matter. The other waste (ashes, etc.) is useful as fertilizer. The other way of utilizing solar energy held in the vegetation is by producing wood alcohol, methanol. Methanol is a clean-burning fluid, composed of two molecules of hydrogen bound to a carbon monoxide molecule. It could be stored, transported, and used much like gasoline, or it could be used to generate electricity in fuel cells. Manufacturing methanol is an old and well-established technology dating back to the 19th century. Its wide-scale use was prevented by the easy availability of oil and natural gas; it was not competitive financially, except in wartime Europe when methanol was used in place of gasoline. With the increasing cost of crude oil, methanol may stage a comeback.

CHAPTER 9

The Energy of Running Waters

The energy from the sun causes enormous quantities of water to evaporate from the surface of the oceans. About 600 calories are required to evaporate one gram of water; $40,000 \times 10^{12}$ watts of solar energy—23% of the whole—is used for this purpose. All the water that is evaporated by this immense energy is carried in the atmosphere, in vapor form, precipitated into clouds. From the clouds carried by winds, the water falls onto the Earth's surface. The raindrops on the land have a potential energy with respect to the level of the sea. The potential energy is the product of the weight of the water and its elevation over the level of the sea. Most of this energy is used up while the water is streaming toward the sea. The energy of the flowing water is lost through friction between the water molecules and between the rocks and the water. Did you ever wonder about the forces that carved out the Grand Canyon, the power that carried all that silt onto the Mississippi Delta? It was the potential energy of the water droplets on their way back to the sea.

If a dam is built in a portion of a stream, the water is forced to rise to the top of the dam before it is allowed to continue its flow downward. A lake is formed in this way in front of the dam, causing the water to flow slower in the enlarged channel without losing much of its energy. The stored water in the reservoir behind the dam contains a potential energy that can readily be utilized by allowing the water to flow through the dam, through an efficient modern water wheel, called a hydraulic turbine. As an example of such installations the power house of Egypt's Aswan Dam is shown in Fig. 9.1.

The energy delivered by a turbine depends on the weight of water that

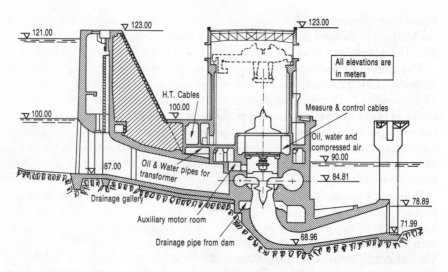

Fig. 9.1 Cross section of the hydraulic power-generating plant of the Aswan Dam on the Nile River in Egypt. (After Ref. 82, courtesy of the Hungarian Academy of Sciences, Budapest, Hungary.)

flows through it in a second, and the height of the dam. Measuring the weight of the flowing water as kilogram per second and the height of the dam in meters, the energy available is measured in meter-kilograms—427 meter-kilograms amounts to one calorie; one kilowatt-hour of power is 367,000 meter-kilograms. It makes little different from a theoretical standpoint whether one kilogram of water falls 100 meters or 100 kilograms of water fall one meter. Hence we can generate just as much power from a small mountain creek falling great heights over a tall dam than a large river of the plains falling a few feet over a relatively low dam. From the standpoint of construction costs, the tall dam with the little water has its advantages, of course.

Water power was utilized for several centuries. Water wheels, driving mills, and saws were common in the Middle Ages. As a matter of fact, even the ancient Babylonians built crude horizontal wheels to harness some of the power of the flowing water. But only the generation of electricity by hydraulic turbines made water power an important contributor of energy. The first such project was built in 1876 for a royal castle in Linderhof, Bavaria, supplying DC power to illuminate a famous cave and a nearby kiosk using arc lights. In 1891 the first high-voltage electric transmission line was put into operation between Lauffen and

Frankfurt, Germany; this signaled the first efficient urban use of hydroelectricity. Before that time the main disadvantage of water-powered electric generators was their great distance from populated areas.

The potential harnessable water power of the world was measured by detailed statistical analysis. The question has arisen in regard to the term "harnessable." During the past century we have found a technological trend to build hydroelectric power stations at sites considered unutilizable for such purposes a few decades earlier. Growth of construction technology, increased sophistication of the nationwide electric power distribution system, and other factors have made considerable changes in the term "harnessable."

Way back, water power was utilized in a river to the point at which the power could be generated at least 80% of the time. Now, we build power stations on rivers to a point at which we know we can count on the river to turn our turbines 50% of the time or less. The reason for this is that today we can count on power stations using coal or nuclear energy to supply us when the river is low. In turn, we can count on a much larger production of power from the water when the rains come because the capacity is there.

While hydraulic power stations are expensive to build, the electricity supplied by them is very cheap. To compare the cost of kilowatt-hours from water power versus other conventional power stations we are somewhat at a loss. In a modern water power development scheme we find other incalculable benefits. How do you evaluate the benefits of a dam holding much of the flood waters that otherwise inundate the agricultural lands below? How to figure the benefits of the stored water when it is used for water supply to cities and irrigation of parched lands during drought? How would you value the recreational advantages of a reservoir? What about the abatement of industrial and municipal water pollution by controlling the flow of a minimal necessary river discharge? There are many advantages to a multipurpose dam project besides the kilowatt-hours produced by its power station. Electricity from water power is considerably cheaper than from fossil fuel plants. It is clean energy and does not cause any pollution as other conventional sources do. It does not add to the thermal pollution of the atmosphere because it uses the sun's incoming energy, which otherwise would be used up as the friction heat in the streaming rivers. It uses the continuously flowing solar energy, saving the stored energy "capital" of the Earth. Hence, water power is a very desirable form of energy that deserves to be used to its utmost.

The world's energy potential from rivers was estimated in a very reliable manner to be 5609×10^6 kilowatts (82). This amount is shared in the various continents as follows:

North America	717.15×10^6 kilowatts
South America	1110.4×10^6 kilowatts
Europe	119.7×10^6 kilowatts
Asia	2308.5×10^6 kilowatts
Africa	1153.6×10^6 kilowatts
Australasia	119.0×10^6 kilowatts

Only about 8.5% of this energy is utilized at present. About 30% of the world's electric energy production comes from hydraulic power. This amount ranges widely among the various countries of the world. The greatest percentage contribution of water power to the total electric production is that of Norway with 99.5%. Next is Switzerland with 98.9%. In these countries water in its most utilizable form is obviously abundant. In the United States the contribution of water power to the total energy utilization is only about 4%; for electric power generation only, it may be put to be about 5.3%. In Europe, by comparison, the contribution of water is well over 30%. The Federal Power Commission estimates that the total utilizable hydraulic power in the United States is about 147,000 megawatts, of which about 33,000 megawatts are already installed. The total ultimate electric output from United States waters could be about 720 billion kilowatt-hours per year.

One of the advantages of hydroenergy production is its high efficiency. While over 60% of the fossil fuel energy is lost as waste heat in a thermal power plant, with hydraulic turbines the conversion efficiency is about 90%—a remarkable figure (82).

Significant growth of water power development in the United States is not to be expected. The best sites for the development of water power are in the western part of the country, while the most intensive use of electricity is at the East Coast. Transportation of electric energy through such long transmission lines is wasteful and hardly feasible at present. This situation may change by the development of underground cryogenic transmission lines. Environmental concern over flooded valleys prevent effective development. The peculiar conditions of our economy and politics prevent large-scale government investment in water power development. As long as "your investor-owned power utilities" as well as powerful oil, gas, and coal lobbies can effectively fight these projects, they will not materialize. Water power development requires enormous investments that cannot be financed realistically by private concerns. On the other hand, the trend is to build large-scale projects for economy. In this matter the political and economic conditions of the United States are a great disadvantage for water power. The following statistics on large projects, and their locations, support these contentions.

The highest capacity hydroelectric plants with the greatest reservoir volumes are as follows (82):

Krasnoyarks, USSR	6000 megawatts
Bratsk, USSR	3600 megawatts
Volgograd, USSR	2530 megawatts
Kuibyshev, USSR	2300 megawatts
Grand Coulee, US	1974 megawatts

Highest dams of the world are:

Laures, Italy	2030 meters
Reisseck-Kreuzeck, Austria	1771 meters
Chandoline, Switzerland	1750 meters
Fully, Switzerland	1645 meters
Juan Carosio, Peru	1440 meters
Portillon d'Olle, France	1400 meters

Our Hoover Dam, by comparison, is only 221 meters high.

As to the capacity of the hydraulic turbines, the biggest are at the following projects:

Krasnoyarks, USSR	510 megawatts
Bratsk, USSR	225 megawatts
Stornorrforsen, Sweden	140 megawatts
Grand Coulee, US	120 megawatts

In light of the tremendous technological superiority of the United States, these above figures show that water power development has been relegated to an inferior position in this country.

One example of the difficulty proponents of water power face in the United States is the case of the proposed Dickey-Lincoln School hydroelectric project in Maine. This plan would call for the building of the world's eleventh largest earth dam, nearly two miles long, to create a 138-square-mile lake in the valley of the Saint John river—a flood-prone stream through the last remaining wilderness area on the East Coast. The size of this project would exceed Egypt's Aswan Dam. Once developed, it would supply electricity during a one-and-one-half-hour peak demand period each day throughout the New England Region. Although the US

Congress authorized the Corps of Engineers to proceed with the project in 1965, it has never appropriated money for it, mainly due to the vigorous lobbying efforts by a strange alliance of environmentalist groups and coal and timber interests.

One of the problems with hydropower is that the power is there when the water is flowing. To store the power of the running waters for periods of peak daily demand, storage reservoirs are built, where possible, at high elevations. During periods of low demand, the electricity is used to pump water up to these reservoirs. From there, the stored water is released through the turbines during peak demand. This scheme adds about 10% to the total capacity of the power station and conserves the utilizable water as well.

Another source of hydraulic power is the fluctuating water level due to tidal motion of the oceans. The world's tidal energy is about 3×10^9 kilowatts. This is a rather small amount compared to solar energy. This is really a utilization of the attractive force of the moon. At some sites (e.g., the Canadian Maritime Provinces and the French Atlantic coast), the tidal motion is considerable. Tidal energy is harnessed by filling and emptying of a bay or estuary that can be closed by a dam. The enclosed basin (chamber) is allowed to fill and empty only during brief periods at high and low tides in order to develop as much power as possible.

Many people still remember the Passamaquoddy project proposed in the 1930s. This proposal (which has been discussed since 1919) in the Bay of Fundy between Canada and Maine would provide 1000 megawatts from turbines installed in a dam between the bay and the ocean, by capitalizing on the intermittent difference of water levels during tides (an average of 18 feet). The problem with such projects is the fact that the power would be generated for only a few hours each day, when the water level differences between the bay and the open sea is the greatest. The cost is another problem. The Bay of Fundy project would take about 10 years to build and would cost about one billion dollars.

Tidal power is already utilized in France at the recently completed plant on the Rance River estuary. Tides there average about 27 feet. The flow into the estuary reaches 18,000 cubic meters per second during the highest tides. Modern, bulb-type turbines generate electricity when the flow from the sea to the estuary and from the estuary to the sea is the greatest. During low flow periods, the turbines are utilized as pumps to increase the water level in the estuary, extending the period of electricity generation. The pumping system results in an increase of overall electric production by more than 50%. The Rance tidal power station produces 550 megawatts, 10% of which is used up by pumping (105).

The world's second tidal power station was put into operation in 1970 near the Arctic port of Murmansk in the USSR. On the Soviet arctic coasts the tides range between 30 and 45 feet and at this particular site conditions for development were ideal. Much larger tidal plants are being designed in the USSR. The next one to be built at Mezen Bay will have a capacity of 6 million kilowatts. An even larger one is envisoned at Penginsk Bay, with a potential capacity of 20 million kilowatts.

But how much tidal power could we produce if all favorable bays and inlets were developed on the United States coast? A somewhat uncertain guess is that it would be somewhere about 100,000 megawatts. This obviously would not be a significant amount in light of the probable cost of such serious construction projects. We would need a great deal more energy in the future, so we have to look for a more satisfactory solution. After all, the world's tidal power amounts to about 2% of the total water power.

Another manifestation of energy in water is the motion of waves. Wave motion is caused by winds. Sustained winds of, say, 80 mph velocity create waves of 50 feet or more on the open sea. Harnessing this large though intermittent source of energy has been proposed frequently. No technical solutions have been offered as yet. One promising study by the Scripps Institute of Oceanography involved a wave pump that produced about one kilowatt of electricity. The device has only one moving part and consists of a 200-foot plastic pipe with a check valve through which the waves force the water. By the action of the waves, the water above the check valve is forced along into a pressure tank aboard a ship or a buoy. Then it is discharged through a turbine to generate electricity. Storage of the generated electricity presents the same problem as windmills, solar collectors, or other intermittent energy producers. But the idea warrants further study as a reliable and cheap electrical energy producer for small, remote locations.

The energy of the sun's evaporating power can be utilized in a rather direct manner at suitable sites of seaboards of hot tropical climates. This scheme is the so-called "depression power plant" or "heliohydroelectric" plant (69). The fundamental principle of operating such a plant is as follows: In hot climates the water evaporates at a rate of two or three meters per year or more. If areas are found on the land surface that are deeper than the level of the sea, water may be conducted into these depressions through channels. Once such a depression is filled with water, the process of evaporation will maintain a constant flow of water in the channel. This flow from the sea level to the depression may be harnessed by turbines.

One such site studied frequently for possible development is the Qattara, located south of El Alamein between Egypt and Libya (12). This place is a desolate desert about 80 kilometers south of the Mediterranean, a vast depression 300 kilometers long and 150 kilometers wide. The deepest point of this desert is 135 meters below sea level. The yearly rate of evaporation is about 1.8 meters. The latest proposal calls for a channel made by underground nuclear blasting. The water, once allowed into the depression, would form a lake with a 12,000-square-kilometer surface area. To balance the yearly evaporation, at least 650 cubic meters per second could be fed continuously into the lake. This flow could produce a peak electric load of 4000 megawatts if the scheme would also include a secondary pumped storage reservoir. In contrast to the energy resource of the Nile harnessed by the large Aswan Dam, where only a portion of the installed 2100-megawatt capacity can be utilized, the Qattara project would supply a continous flow of energy, independent of seasonal fluctuations. The presence of the big lake in the desert would be of beneficial influence to the weather, helping to reclaim vast areas that are today thoroughly useless. The recently discovered oil and gas of the Libyan desert could be exported instead of used locally, since the produced electricity would be sufficient to fulfill domestic demand. The navigable lake would provide cheap transportation from chemical plants proposed near the oil, reaching the open sea through the channel. Compared to the local oil and gas, the electric energy would be inexhaustible from the turbines of the Qattara.

Similar land-based evaporation power schemes may be built at the Dead Sea or perhaps in the Turkmen Desert near the Caspian Sea. Even California's Death Valley, 85 meters below sea level, has some potential for depression power plant development.

Other potential land-based depression power plants (82) were proposed at the following sites:

Lake Assal, Somalia	67 million kilowatt-hours
Lake Assale, Ethiopia	540 million-kilowatt-hours
Chotts Melrhir, Algeria	270 million-kilowatt-hours

Much larger depression-type projects could be built in bays or gulfs of the open seas. For example, the western part of the Persian Gulf can be transformed into a closed reservoir by building a levee from Saudi Arabia to the island of Bahrein and from there to Quatar. This is the so-called Dawhat Salwah area. The yearly evaporation in this region is about 3.5

meters. The proposed reservoir area would be 2300 square kilometers. The water elevation difference would be about 14 meters, after a few years of initial evaporation. The Dawhat project would provide a constant rate of electric energy production about six times that of the Qattara project—about 300 million kilowatt-hours per year. A similar depression power plant could be built at Bab el Mandeb, closing the Red Sea. This could have an annual energy output of 105,000 million kilowatt-hours. Another potential site is the Maracaibo Bay in Venezuela. Similar proposals were made about closing the Straits of Gibraltar. The evaporation rate of the Mediterranean Sea is larger than the inflow of the surface waters and rain waters into it. To keep the level constant in the sea, there is a stream of water flowing into the Mediterranean at Gibraltar. Such projects may seem utopian today, but some of them may well become realistic in the years to come.

An entirely different concept of energy utilization from water is based on the temperature differences in nearby water bodies. This is what is called "hydrothermal" energy. Since 1933, a French built experimental power station yielding 11 megawatts of energy has been in operation at Abidjan on the Ivory Coast. It converts the energy inherent in the 20 to 25° Celsius difference between warm coastal lagoons and the colder open sea.

Ocean thermal gradients are another prospect. Due to the absorption of sunlight in the upper layer of the oceans, combined with density separation by depth, a vertical temperature profile is established in the water of the oceans. These differences are enhanced at certain areas through the action of ocean currents. Large amounts of solar energy are absorbed in the equatorial regions of the oceans. One resulting current of warm water, the Gulf Stream, flows northward along the eastern coast of the United States and a countercurrent of cold waters flows southward beneath it. In some locations (for example, 30 miles off the coast of Miami, Florida, and in the Caribbean Sea) the warm and cold streams are separated vertically by about 2000 feet. The temperature difference between these two streams remains essentially fixed between 35° to 45° Fahrenheit. A collection system of units moored at one mile spacing along the length and breadth of the Gulf Stream off Florida's coast is thought to be capable of energy production equal to the total consumption of the United States in 1970. A conceptual design for such an "ocean thermal gradient" power plant was made and it has been estimated that such units could produce electric power at competitive cost. The concept involves a vertical pipe 38 feet in diameter suspended to 2000 feet below sea level to tap the cold water source. Using a low temperature vapor cycle turbine,* utilizing propane as heat exchanger fluid, such a unit could generate 1000 megawatts of energy. Proponents of this system claim that one percent of electricity used by the United States by the year 2000 could be provided from ocean temperature differences. Skeptics point out that the efficiency of the system would be very low, at about 2% because of the low temperature differences involved. The cost of such a project at a 100-megawatt rating would be about $100 million. The location is noted for its hurricanes, making the safety and reliability of the seaborn power stations rather dubious. The low temperature vapor cycle turbine system is yet to be developed. When it will become available, thermal energy from the Earth would be a more abundant source than ocean thermal differences, at considerably less cost. All in all, the future of these offshore proposals does not seem very rosy.

Water power generation may well become much more important in the future when the separation of hydrogen to feed fuel cells** becomes a commercial reality (21). Then water's energy would probably prove to be the cheapest of all power generators.

*This was discussed in Chapter 7.
**More on fuel cells in Chapter 13.

CHAPTER 10

To Catch the Wind

Windmills were one of the earliest forms of energy utilization by mankind. They came into prevalent use in Europe during the 12th century. Wind-driven pumps helped dewater Holland's coastal areas through her early canals. Flour was generally produced by the aid of millstones driven by windmills all over Europe (22). Windmills were used in Persia and China thousands of years before Christ, and in 1700 BC the Babylonians used them to pump water for irrigation. To use windmills to grind corn was a 15th-century Dutch idea. Using windmills for sawing was introduced much later.

Remnants of an ancient wind generator predating the earliest ones in Europe were found in the border area between Iran and Afghanistan. This one operated with wheels on a vertical shaft, with one side shielded from the wind by a brick wall, the other exposed to the wind (19).

Even before winds were used on land, sailing vessels replaced oar-driven galleons. The development of fossil fuel technology, steam and internal combustion engines replaced man's reliance on the capricious winds. Today only a few windmills of the once 6.5 million dotting the American midwest's farmlands remain in use, raising water from wells.

Yet the energy in the winds is considerable. As one of the manifestations of the sun's energy's uneven distribution of heat on the Earth, winds indicate the movement of hot and cold air masses over the surface. On some areas of the Earth, winds are more prevalent than on others. The reason for the existence of such windy sites is complex.

Wind power available throughout the Earth is believed to be equivalent to 100 billion watts per year. Winds are stronger at sea than over the land surface. In 1939 Sverre Peterssen analyzed the wind power potential of the oceans. His results are shown in Fig. 10.1. The numbered lines in this map represent estimates of the probable yearly production of kilowatt-hours by each kilowatt of installed wind-generator capacity. For example, a 1000-kilowatt wind generator installed at a point along a line marked 3000 could provide 3 million kilowatt-hours in one year.

In the United States it is estimated that up to one trillion kilowatt-hours of electricity could be generated in a year—more than half of the electric production of the country. This energy is utilized not only in moving air currents but in driving up waves, eroding land surfaces and in many similar ways until it finally is dissipated due to friction, creating a corresponding rise of heat content in the atmosphere. The average wind power measured in the Oklahoma City area was found to be about 18.5 watts per square foot of area perpendicular to the wind's direction. This is roughly equivalent to the amount of solar energy received by a square foot of land in Oklahoma, if one averages the sunlight as if received over day and night in all weather conditions.

A 100-kilowatt-capacity wind generator built in 1931 at the Crimean Balaclava in the USSR was one of the first in commercial production. Scandinavian countries, utilizing windy coastal areas, gave considerable attention to wind-produced electric power, due to the fact that they are notably short of fossil fuel. In particular, Denmark produced 3 million kilowatt-hours of electricity in 1944. Since that time additional new, large-capacity wind generators were installed. Presently (1974), the world's largest wind generator is located in Celinograd in the USSR, generating 1200 kilowatts of electricity. To supply electricity in remote areas, more than 20,000 wind generators are being installed every year in the Soviet Union alone. During World War II the world's largest wind generator operated near Rutland, Vermont. The machine was 110 feet high. Its two 8-ton blade propellers had a 175-foot diameter. The generator produced 1500 kilowatts of electricity—enough to light the town. After three and one-half years of operation, one of the blades broke off, ending the experiment.

To convert the wind's energy into mechanical energy—and subsequently into electricity—is relatively easy. But of all natural energy sources, wind is the most capricious. Wind power at suitable windy sites could be used and may play an important role in the development of an individual area, but for large-scale development of electricity it will, in all probability, remain a minor contender. Even if extensive utilization of wind were technically feasible, the danger of possible modification of our weather would retard its full-scale application.

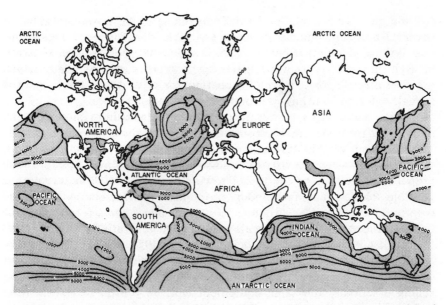

Fig. 10.1 Map showing the wind power potential over the oceans. (Courtesy of Prof. William E. Heronemus, University of Massachusetts, published in *Oceanus*, Vol. XVII, Summer 1974, by the Woods Hole Oceanographic Institution, Woods Hole, Mass.)

Like other intermittent energy sources, wind use will be more desirable if an effective method of large-scale energy storage could be developed. Such promise may be realized through the development of efficient fuel cells.* One scheme that seems to offer a solution after successful research is to use the variable power output of the wind-driven generator to decompose water into hydrogen and oxygen. These would be stored under pressure and recombined in a fuel cell to generate electricity on a steady basis.** An alternative solution is to ship the compressed hydrogen to other sites where it may be burned in a gas turbine.

With the design of more efficient aeroturbines, the interest in using wind energy is reawakening. A proposal was made recently to building wind generators on floating platforms in the Atlantic off the coast of New England. These would use the gained wind energy to decompose sea water and produce hydrogen for fuel cells. A group of such aeroturbines could produce the equivalent of 360 billion kilowatt-hours of electricity a year.

*More on fuel cells in Chapter 13.
**More on hydrogen in Chapter 13.

Wind generators are commercially available in Australia as well as in several European countries (France, Switzerland, and others). The cost of one of these products is between $2000 and $5000. On the German island of Sylt, a novel-type wind generator was recently installed, which provides electric power (including heating) for several families. Instead of a single rotor this one has two rotating in opposite directions. To eliminate otherwise necessary gearing, the electric generator is located between them. The rotor design incorporates advanced aerodynamic concepts. The efficiency of this device is more than 60%.

Technology already exists to build wind generators with 200-foot long propeller blades set on the top of towers as high as 1000 feet. Another possible solution recommended by proponents of wind energy utilization is to build a rectangular structure to support many small rotor-generator units. At the other end of the scale a small wind generating unit was put on the market by Hughes & Associates Inc. in 1974 to operate a battery recharger on offshore drilling platforms. Of the score of windmill manufacturers offering their wares 90 years ago, three are still in business selling essentially the same equipment with very little change in their design. These units are so well engineered that they require virtually no maintenance and almost none of their parts wear out.

Soviet standard models, called Chayka, are made in various sizes, with generator capacities ranging between one and 15 kilowatts. This model has 12 propeller blades and is supported on a three-legged steel stand. These are used mainly for pumping. The also Soviet-made Buran is a three-blade propeller model made for driving a pump directly (19), lifting water from depths of 150 feet.

A much more modest wind generator was recently demonstrated by Canadian scientists. It consists of two or three convex metal blades of aerofoil design, attached to a vertical shaft that is supported by wall bearing at top and bottom and held by guy wires at the top. The machine has the appearance of an elongated kitchen blender. This device, having a vertical axis of rotation, can utilize winds from every direction without adjustment of orientation. As the wind strikes the blades, they turn around the vertical shaft. At a wind speed of, say, 15 mph the blades revolve around the axis at 170 rpm. At this speed, the 15-foot diameter turbine is said to produce 0.9 kilowatts of electricity. Machines like this could potentially be helpful in underdeveloped areas, provided that a cheap method is found to store electricity.

Wind research is reawakening in the United States due to the recent oil crisis. The National Aeronautics and Space Administration is currently

(1974) designing an experimental 100-kilowatt wind generator at a cost of $200,000 with propellers of 60 feet in length, supported by a 100-foot tall tower. This generator will be installed at Sandusky, Ohio. Even larger ones are being planned, at a capacity of 1000 kilowatts, enough to supply electricity to 200 households at wind speeds of 20 miles per hour.

Canadian researchers propose the utilization of wind energy for the heating of homes by having windmills drive heat pumps or other heat-generating devices. Presently, the main problem in this development appears to be the lack of an efficient method to store heat.

Wind may also stage a comeback as a mover of ships. German engineers were completing plans in 1973 for a 400-foot, 17,000-ton freighter whose square sails would be set, reefed, and furled by a computer-aided machine controlled by a single man in the pilot house. The sails would take the place of the large diesel engines and are expected to be able to give the ship a speed of 12 to 20 knots, well in the range of motor-powered freighters. The ship would have an auxiliary engine and would utilize modern satellite-based navigational aids to seek out the best winds. Being fully automated, such ships need much smaller crews than conventional ships, designers claim.

Many aspects of wind energy utilization are studied in engineering laboratories around the world. Such activities have rapidly increased in recent years. Research is coordinated by the Wind Engineering Research Council funded by the National Science Foundation. Recent reports by a panel of specialists estimated that the maximum total energy that could be generated by wind power would be about 1% of the predicted annual electricity consumption of the year 2000. By 2020, this may rise to 5.2% worldwide, if the necessary research investments would be made.

It is doubtful that government and industry would be willing to support more significant research in the area of wind power utilization in light of the fact that this resource could affect the total energy picture at only a marginal level. But it is entirely possible that wind energy will be a serious contender in electric supply at remote locations, along with solar heating and fuel cells.

CHAPTER 11

Atomic Energy

There is considerable confusion in the public's mind when it comes to nuclear power plants. There are the proponents, who consider it to be the ultimate salvation of the world's energy needs (18). There are the opponents, who foretell the doom of mankind either by a catastrophic accident or by slow death due to ever-increasing environmental pollution. In either case our government is heavily committed to the development of nuclear power to face the rapidly increasing energy demands of the future. Whether one likes it or not, atomic energy is proven to be feasible from both scientific and commercial standpoints and it is here to stay (71).

A nuclear power plant is very much like a conventional steam power plant. The only difference is that the heat used to heat the steam that runs the steam turbines (67) and hence the electric generator is not obtained by burning coal, gas, or oil but is derived from controlled nuclear reactions (9).

What is a nuclear reaction? Natural radioactivity was discovered at the end of the 19th century. This is the property of some substances to give off rays without external stimulus. These rays were known from cathode rays in vacuum tubes, invisible but penetrating X-rays and phosphorescent substances. Soon scientists found that the radioactivity of the element uranium involved a decomposition of its atomic nuclei. In the process the nuclei of the atoms are altered and the identity of the element changed. The spontaneous decomposition of the uranium results in thorium and helium, whose combined atomic numbers (234 + 4) equal that of uranium. There are many successive steps in this decomposition, each involving intermediate isotopes that are unstable and decompose further.

111

During this process of decomposition, the substances give off rays of small particles of high velocity, and gamma rays that are a form of penetrating invisible light rays. After a long period of time, the final, stable product of this decomposition is lead. The time required to decompose about 99% of uranium into lead is over 22 billion years. The time of decomposition of the various radioactive isotopes is expressed by their "half-life." This is the time required to decompose one-half of the amount of the isotope. For example, the half-life of thorium is 24.5 days, the half-life of radium is 1590 years, and so on. In 1938, Otto Hahn and other German scientists found that a collision with a subatomic particle (called a neutron) may cause the uranium nucleus to split into other isotopes. Soon after, on December 2, 1942, three scientists of the University of Chicago—the Italian Enrico Fermi, and Hungarians Leo Szilard and Eugene Wigner—built a device to show that such nuclear splitting, called fission, can be maintained in a controlled, self-sustaining chain reaction. This device was the first nuclear reactor.

The subatomic particles composing an element are held together by a very large binding energy. Once an element is split up into pieces, some of this energy is released. The fission of uranium-235 releases about 2×10^{10} calories of heat. Thus, one pound of uranium will generate as much heat by fission as the burning of 1250 tons of coal, which is equivalent to 11 million kilowatt-hours of heat. One watt of thermal power corresponds to the splitting of 30 billion uranium atoms per second.

Natural uranium is composed of 0.7% uranium-235 (U-235) and 99.3% of uranium-238 (U-238). These isotopes are chemically identical but exhibit different nuclear properties. U-235 sustains a chain reaction if assembled in a relatively small amount—the size of a football, for example. This minimum amount to maintain the chain reaction is called "critical mass." U-238, on the other hand is incapable of sustaining a chain reaction (56).

A modern nuclear reactor may contain over 100 tons of uranium fuel. The fuel loading of a 100-megawatt reactor contains enough fuel to equal the thermal energy of 20 million tons of coal. If this amount of coal were available for a conventional coal-fired plant, it could operate for eight years without additional coal shipments. Part of this reactor fuel is removed and replaced with fresh fuel each year (60).

Modern reactors use uranium pellets that are enriched to contain more U-235 as fuel. The enrichment is done in special gaseous diffusion plants. These plants convert uranium ore into uranium hexafluoride gas, which is passed through a series of porous membranes. These selective barriers concentrate the U-235 content of the gas. Operation of these gaseous diffusion plants for uranium enrichment requires a considerable amount of energy. The process is expensive but the resulting fuel is well worth the effort. Most of the enrichment plants are located in the United States; only about 5% of the world output comes from foreign sources. The American uranium enrichment technology is in a leading position in the world. Output of enriched uranium fuel is expected to rise from 17.5 million kilograms in 1975 to 72.5 million kilograms in 1985. In the United States alone, $270 million was spent on enrichment in 1972. By 1976, this amount will rise to $340 million.

In late 1973 France announced the impending construction of a $1.7 billion gaseous diffusion plant to be built in consortium with several other European countries. Other European countries would prefer the use of the centrifugal enrichment method, which is cheaper because it does not require the large amounts of electric energy needed in gaseous diffusion. The cost of enrichment is about 40% of the cost of nuclear fuel preparation.

Uranium enrichment is a costly operation from an energy standpoint. In 1974 the existing three enrichment plants consumed 3% of the electricity produced in the United States. But in the future, experts claim, every kilowatt used for enrichment will result in 30 kilowatts of electricity produced in nuclear plants.

Supplying the fuel for nuclear power plants involves a series of operations. These are as follows:

Mining of uranium ore. A typical high-grade ore contains three to six pounds of uranium per ton.

Milling the ore and separating the crude uranium concentrate.

Refining the concentrate by removing impurities, and converting it to uranium hexafluoride gas.

Enrichment by gaseous diffusion to increase its U-235 content. Chemical conversion into uranium dioxide form.

Fabrication of fuel assemblies by forming the fuel pellets, loading them into tubes and subassemblies.

After use in a reactor, some of the fuel rods are lifted out of the core under the protection of water. These rods are allowed to "cool" in a huge concrete container. After cooling, they are shipped in lead containers to fuel reprocessing centers. There the spent fuel is separated into plutonium—a by-product of the reactor, and long-lived fission products such as cesium-137 and strontium-90. The latter two are converted into solids and stored for shipment to a Federal Radioactive Waste Depository. Waste disposal is still a major problem of the nuclear industry (40). Efforts to store the waste underground have not been fully successful so far. Mismanagement of these highly radioactive wastes threatens extensive areas of the United States with massive contamination. More than one-half million gallons of deadly liquid has leaked from storage tanks near Richland, Washington (4). Other leaks at a similar facility at Savannah River, South Carolina, are threatening the ground water supplies. Similar traces were found near Idaho Falls, Idaho, threatening the water supplies for much of the Pacific Northwest (94). The nation's first water supply system contamination by radioactive wastes was reported in 1973 from Broomfield, Colorado (95). Tritium, a radioactive isotope of hydrogen, was found in a surface water reservoir in concentrations ten times the normal background level. It was also found in the urine of the town's residents. The reservoir was contaminated through ground water from buried wastes of a nuclear weapons factory five miles away. The accumulation of plutonium in a trench at the Richland Reservation has reached a level at which nuclear chain reaction is possible. Meanwhile the radioactive wastes from American-built reactors in Japan, Canada, and Italy are imported to the United States through our commitments to reprocess these fuels (47). Some scientists feel that such problems have brought the country to the brink of national emergency.

Cost of the nuclear fuel preparation cycle to keep our reactors working in the United States amounted to the following expenditures in 1972:

fabrication	$184 million
enrichment	220 million
uranium oxide conversion	25 million
uranium oxide supply	98 million
reprocessing	5 million
TOTAL	$532 million

The United States Atomic Energy Commission estimates that the uranium oxide requirement during the decade of the 1970s will be 206,000 metric tons. In the same period the world's need (excluding China and the USSR) will be 430,000 tons. Presently, there is a short-term oversupply of uranium ores in the United States, caused largely by reactor construction delays. This brought about depressed prices and a decrease of exploration activity. The all-time high in exploration drilling was 1969, when 76,000 bore holes were drilled in search of uranium deposits. In that year over 27 million acres of land were held for uranium exploration and development. Ninety percent of the American uranium drilling activity was concentrated in the states of Wyoming, Texas, and New Mexico. There is considerable secrecy shown by uranium exploring companies about their deposits, although they release their data on a confidential basis to the Atomic Energy Commission. The United States domestic uranium oxide (U_3O_8) resources in 1971 were put at 736,000 tons producible at $8 per ton, plus an additional 980,000 tons producible at $10 per ton. Past uranium finding rates kept pace with drilling rates reasonably well, indicating that significant additional reserves could be found in the future, at least until the breeder reactor program becomes commercially feasible. Fig. 5.1 shows the locations of the uranium fields in the United States.

New supplies are continually found around the world. For example, at Jim Creek in the Northern Territory of Australia, more than 5 million tons of ore reserves contain some 24,000 tons of uranium oxide. But the U-235 supply of the world would not be able to provide energy with the present technology for much time after the year 2000.

Present-day economics do not warrant mining low-grade uranium ores. But in the future these relatively abundant supplies certainly will be utilized. One such example is the Chattanooga black shale, which is found at mineable depths in much of Tennessee, Kentucky, Ohio, Indiana, and Illinois. The uranium content of this shale is about 60 grams per metric ton ($6 \times 10^{-5}\%$). The average thickness of the layer being about five meters, the energy content per square meter of surface area is equivalent to about 2000 tons of coal or 10,000 barrels of oil. The energy held in this Chattanooga shale represents an immensely large supply in the form of low-grade nuclear fuel. Many similar sources have been located around the world.

Recent research studies indicate some possibility of using boron as an atomic fuel in the distant future. Boron, a light element, may undergo fission and release energy without producing polluting radioactive by-products of long half-life. Boron is abundant on the Earth, in the ocean and in dry lake beds as borax. Boron, and perhaps other elements that could be made fissionable in the future, may make the future of the nuclear industry even brighter.

The reactor core in which the fuel is placed is a thick-walled container to prevent radiation leaking into the environment. Within this heavy shield, one finds the fuel emplaced in thousands of 12-foot-long tubes. Assembled in 7×7, 8×8, or 15×15 arrays, these tubes are made of zirconium alloy and hold the uranium fuel in the form of small pellets. Figure 11.1 shows a uranium pellet.

Fig. 11.1   Uranium fuel pellet. (An EXXON photo, courtesy of Exxon Corporation, New York, N.Y.)

The fuel rods in the reactor are immersed in a coolant, which may be water, gas, oil, sodium, or another suitable liquid. The purpose of this coolant is to slow down the velocity of the neutrons because they produce more fission at slower speeds. Also the coolant serves as a heat transfer medium to carry the heat outside of the reactor core and transfer it to steam through heat exchangers. Because of the high rate of fission inside the fuel rods, the temperature there is in the range of 4000°F. The coolant assures that the reactor core does not overheat, which could result in melting of the zinc alloy fuel rods. Should they be melted, the fuel would collapse into a pile on the bottom of the reactor core. There, in an uncontrolled fission, it could result in an accident endangering life within 100 miles. To prevent any such happening, a separate back-up coolant system is installed for every reactor.

The operation of a nuclear reactor is provided by an array of cruciform control rods that may be moved in and out of the reactor core between the fuel rods. The control rods contain neutron-absorbing elements such as boron. Once these control rods are inserted fully between the fuel rods, they absorb the neutrons and thereby prevent them from hitting uranium fuel particles, thus initiating fission. As the reactor is started, the rods are slowly withdrawn, allowing fission to develop. The initial supply of neutrons comes from a polonium-beryllium source. At a certain position, the chain reaction will sustain itself, the reactor goes "critical." Further withdrawal of the control rods increases the power output.

In case of accident the reactor core is designed so that the control rods are driven in automatically. This shuts down the reactor by preventing further fission. Of course, it takes some time to have the reactor "cooled," because even after such shut-down the core still contains a large amount of split uranium atoms in addition to the residual heat. After a passage of time, the short half-life isotopes disappear and only the longer lived nuclear products contribute to the heat generation. The radioactivity of a shut-down reactor of 1000-megawatt capacity that has operated for a year has enough residual fission product even after one month of shut-down to equal the radiation of 100 tons of radium. This high radioactivity makes the operation of a reactor a rather tricky job indeed. To assure safe operation, there are a number of back-up systems to monitor and control the reactor. There are several thousand technical standards proposed to be applicable to nuclear reactor design and operation, relating to materials, testing, design, construction, instrumentation, machinery performance, and process control. Only a few hundred of these were passed and approved for use by 1970 (34). The problem is that the fantastic progress in the development of the nuclear industry makes yesterday's standards and specifications obsolete by tomorrow.

There are various main types of nuclear reactors (24, 37). All of these operate more or less on the same principle, and all convert the heat generated by the reactor core into steam. The steam itself is used to turn conventional turbines like those of fossil fuel plants. Some main types of these reactors are as follows:

The boiling water reactor (BWR) uses water as a coolant. The temperature maintained in the core allows the formation of steam. The steam, in turn, runs the steam turbine. This system eliminates the need for a separate heat exchanger. The spent steam is recondensed, cooled, and allowed to reenter the reactor core. A cross section of a BWR plant is shown in Fig. 11.2. Figure 11.3 shows the typical BWR steam system.

In Canada pressurized heavy water (D_2O) is used in most reactors (CANDU-PHW). The heavy water coolant is pumped through the core from where it passes on to a steam-generating heat exchanger. This steam, in turn, drives the turbine.

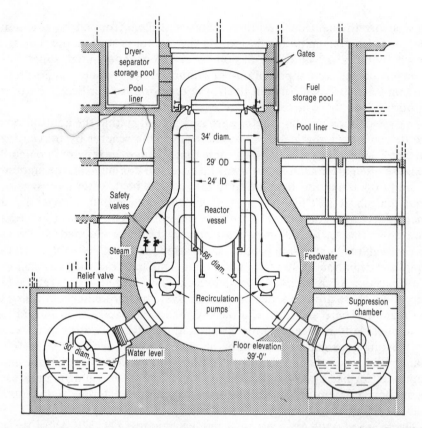

Fig. 11.2a   The arrangement of a BWR reactor core. (After Ref. 71, Copyright © 1973 by *Consulting Engineer*. All rights reserved.)

A variant of the Canadian PHW system is a proposed organic cooled reactor (OCR). In this the heavy water acts as moderator. It passes the heat on to a coolant that is an oil-like substance. This substance has a low-pressure, high-temperature system leading to good thermal efficiency. Both Canadian systems use natural uranium fuel without enrichment (59).

The pressurized water reactor (PWR) is a popular design manufactured by three different American firms. It operates at a coolant pressure of more than twice that of the boiling water reactor. At this pressure, the water is prevented from boiling. The cooling water transfers the heat picked up in the reactor to U-shaped boiler tubes in which water circulates at lower pressures. The resulting heat transfer creates steam in these boiler tubes that drives the turbine.

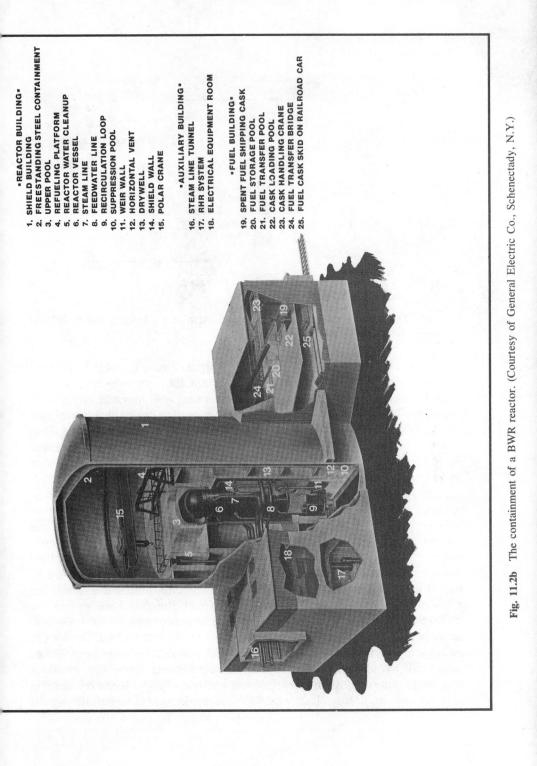

■REACTOR BUILDING■
1. SHIELD BUILDING
2. FREESTANDING STEEL CONTAINMENT
3. UPPER POOL
4. REFUELING PLATFORM
5. REACTOR WATER CLEANUP
6. REACTOR VESSEL
7. STEAM LINE
8. FEEDWATER LINE
9. RECIRCULATION LOOP
10. SUPPRESSION POOL
11. WEIR WALL
12. HORIZONTAL VENT
13. DRYWELL
14. SHIELD WALL
15. POLAR CRANE

■AUXILIARY BUILDING■
16. STEAM LINE TUNNEL
17. RHR SYSTEM
18. ELECTRICAL EQUIPMENT ROOM

■FUEL BUILDING■
19. SPENT FUEL SHIPPING CASK
20. FUEL STORAGE POOL
21. FUEL TRANSFER POOL
22. CASK LOADING POOL
23. CASK HANDLING CRANE
24. FUEL TRANSFER BRIDGE
25. FUEL CASK SKID ON RAILROAD CAR

Fig. 11.2b The containment of a BWR reactor. (Courtesy of General Electric Co., Schenectady, N.Y.)

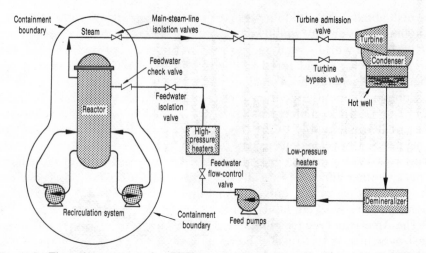

Fig. 11.3   The steam system of a BWR reactor. (After Ref. 71, Copyright © 1973 by *Consulting Engineer*. All rights reserved.)

A new reactor system is the high-temperature gas-cooled (HTGR) reactor. Such plants use helium as a coolant that removes heat from a graphite core. The fuel is enriched uranium and thorium in carbide particles. The thermal efficiency of these reactors is significantly higher (about 40% efficiency) than that of reactors using water as a coolant. This reduces the need for a large cooling water supply. From a thermal disaster standpoint, this system is considered much safer than earlier designs, due to its massive graphite moderator in the core.

A radical improvement of reactor design is represented by the development of the liquid sodium-cooled reactor, the so-called liquid metal fast breeder reactor (LMFBR). This reactor, now being developed in the United States, is capable of producing more nuclear fuel than it consumes. This apparent paradox comes from the fact that the neutron bombardment of the uranium-238 in the fuel results in the forming of unstable U-239, which in turn decomposes into plutonium, itself an efficient nuclear fuel. Successful operation of the fast breeder reactor utilizes 70% of the fission energy of the uranium oxide. By breeding, thorium-232 becomes fissionable uranium-233. With breeding, almost the entire supply of natural uranium and thorium would become available as fuel. This would multiply the energy recoverable from our uranium resources, allowing them to satisfy our electric energy needs for several generations to come. The world's depletable uranium supply would furnish about 300×10^{12} kilowatt-years by breeding, while without it only 1/100th of this energy would be available. The United States shares about 1/10th of the world's such nuclear energy. If successfully developed, the breader reactor would come on line in 1980 at a cost of $750 million. By then a new generation of nuclear reactors will be inaugurated.

Foreign competition to the American breeder reactor program is fierce (2). The USSR completed at least one such reactor of 350-megawatt electrical capacity by 1972. This reactor, built at Shevchenko at the Caspian Sea provides electricity for desalinization of sea water. In the winter of 1974 a major explosion is said to have occurred at this plant. This and other difficulties are allegedly delaying the Soviet LMFB reactor program. Other major industrial nations (such as France, Japan, and Germany) are working intensively on similar developments (59).

The American breeder program appears to be beset by delays and large cost overruns. A test reactor being built in the state of Washington has cost overruns of hundreds of millions of dollars. Critics claim that most of these were due to the unimaginative and cumbersome management by the AEC.

As to the total electric-generating capacity by nuclear reactors in the United States, 84 units are expected to be in operation by 1975. Their total electric power output at that time will be about 65,000 megawatts. Present plans call for the construction of six new plants per year during the coming ten years (126). Plans to construct eight floating nuclear plants off the Atlantic coast were announced in December 1973. The electric capacity of these plants will be 1150 megawatts each. The plans call for building four units 12 miles off the coast at Atlantic City, N. J., two off the Florida coast at Jacksonville, and two near the Mississippi delta. These structures will supposedly be able to withstand hurricane winds up to 156 mph.

At the end of 1971, only 8400 megawatts of electric capacity were in operation of the originally scheduled 23,000 megawatts. The reasons for such delays rest in the lack of standardization due to the many different industrial firms working on these projects, to technical difficulties cropping up before the projects are completed, and to energetic resistance to the construction of nuclear facilities by environmental groups (127). Nuclear power plants have performed with an excellent safety record to this point. But the possibility of an accident looms large in the public mind. The reason for this may be explained by the magnitude of the consequences of a large-scale failure (34). The Atomic Energy Commission estimated that a large accident could result in having 3400 people killed and 43,000 injured. The probability of such an accident is claimed to be in the range of one chance in a billion per year (78). This appears to be a small ratio but happenings of technical failures defied statistical analysis. The common adage of engineers is "if something can happen, it will." The sinking of the "unsinkable" Titanic and the November 9, 1965, "blackout" of the northeastern United States are but a few such examples for totally unexpected technical failures. So far a number of near-misses did happen in the nuclear field. On November 10, 1957, a reactor in Britain failed, covering a 400-square-mile area with fission products. Subsequent investigation showed that all the safety systems had failed simultaneously. A nuclear fuel "meltdown" in Michigan almost resulted in a major disaster. Some opponents of nuclear power claim that such a disaster could eclipse the magnitude of the atomic blast at Hiroshima. From the explosion of a 100 to 200-megawatt reactor, an area in a 15-mile radius would be a deathtrap. Nuclear experts violently oppose such statements. Damage to agricultural products could cover an area as large as the states of Pennsylvania, New York, and New Jersey combined. The possibility of human error is large. Trained manpower to operate nuclear facilities is scarce. Accidents involving nuclear operations were numerous in the past. They ranged from a $200,000 fire in a reactor because electricians accidentally tripped some valves during routine maintenance to vehicular accidents while transporting nuclear materials. Inadequate quality control and improper engineering design procedures have been known to result in serious technical difficulties during the construction of reactor after reactor.

In the fall of 1973, 12 out of 38 nuclear power plants were shut down for repairs and only five of the remainder operated at full capacity. In the summer of 1974 almost half of the existing plants were shut down for repairs of cracks in the cooling systems. The causes were many. Unexplained wear, tear, and cracks in the core, burst steam pipes, dangerous vibrations, and malfunctioning equipment. In all, the nation's nuclear energy producing facilities operated below 60% capacity, indicating a somewhat unreliable performance record for the nuclear industry. Operating difficulties were found in reactors in Wisconsin, New Jersey, Minnesota, New York, Connecticut, Nebraska, and elsewhere. Meanwhile the magnitude of the reactor projects keeps growing. More and more reactors are proposed for locations close to highly populated areas. Major failure of a reactor proposed in the New York City area—at least five are being considered—could wipe out the city. Of 42 reactors in 1969 only two were found to be farther than 30 miles from population centers, making such dangers acute.

Even after barring the possibility of a nuclear accident, the danger of slow nuclear poisoning of the whole population during the coming generations is still very much present. The fact that nuclear facilities are producing a dangerous buildup of radioisotopes in the environment was well documented. There is no safe low level of nuclear radiation dose below which no danger to life exists. Discharges from some plants were almost immeasurably small in water effluents but in the vegetation in such waters the proportion was larger. In animals feeding on such plants the radioactivity was found to be up to 40,000 times greater—the radioactivity of egg yolks of water birds was found to be more than a million times greater. This example shows how the "immeasurably small" water and air pollution of nuclear radiation builds up in nature to a level at which food of any kind could become poison for man (34).

On the other hand, the AEC and the nuclear industry point out the fact that there were only seven radiation-caused deaths in 30 years. They claim that pollution due to nuclear waste was largely related to the nuclear weapons program. Also, new technology is expected to make is possible by 1976 to dispose of nuclear waste in evaporated solid form. Even by the year 2000, the solidified nuclear wastes from some 1000 reactors could be "stacked on a tennis court" (126).

Scientists are hopeful that the danger of accidents or creeping pollution by nuclear reactors can be avoided. The federal government is investing heavily in nuclear technology. The power of the atom will take much of the load off our dwindling fossil fuel resources.

By the year 2000, 1000 nuclear reactors are expected to operate in the United States with an average capacity of 1000 megawatts each. In 1973 there were only 37 reactors completed with a total production capability of 21,687.4 megawatts. In addition some 57 were under construction and 89 more planned. By 1980, some 21% of the nation's electric demand will be supplied by nuclear reactors. By 2000, the nuclear industry is expected to be a mature one, supplying the bulk of our electric needs (126).

CHAPTER 12

Fusion, the Promise of Limitless Power

It fuels the sun. It happens in the hydrogen bomb. And, if we really work on it, perhaps we could harness nature using controlled fusion energy to feed us power for thousands of years to come.

The physical concept of fusion is simple. By creating immensely high pressures and temperatures, light atoms can be combined—fused together—to form a heavier atom. In the process some of the energy that bonded the light atoms together is released. The lightest atom is hydrogen. On the sun, ordinary hydrogen is fused together by violent burning, forming helium (7). The cause of the fusion is the gravitational contraction, which creates the tremendous pressure needed for the reaction. Scientists computed that over ten billion years must pass before all the initially existed hydrogen will burn up on the sun (39). In the case of the hydrogen bomb, two heavy isotopes of hydrogen, called deuterium, are combined to form helium. For each pair of deuterium, one atom of helium-3 is formed and one neutron particle is released along with energy amounting to one million electron volts (1.6×10^{-13} joule) (41).

For controlled fusion, the idea is somewhat similar. Here we attempt to fuse one deuterium atom with one tritium atom (both heavy isotopes of hydrogen) the former containing two neutrons, the latter containing three. The resultant element of this fusion reaction is helium-4, a helium isotope containing four neutrons. For each helium-4 atom, one neutron is also released, plus 17.6 million electron volts of energy (31.36×10^{-3} joules). Typical fusion reactions are compared to fission reactions in Fig. 12.1.

Advantages of fusion rest on the superabundance of deuterium in nature. Hydrogen is a part of water, H_2O. One deuterium atom exists for

125

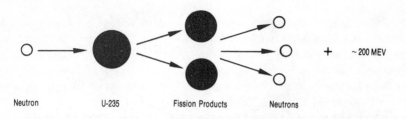

FISSION REACTION

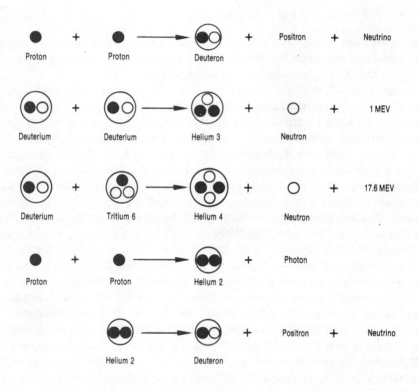

FUSION REACTIONS

Fig. 12.1 Schemes of nuclear fission and fusion reactions.

each 6700 simple hydrogen atoms in water. There are about 10^{25} deuterium atoms in one cubic meter of sea water, amounting to about 34.4 grams. If the fusion energy of the deuterium found in one cubic meter of sea water is released, it can generate 7.94×10^{12} joules of energy. This is equivalent to the heat of combustion of 300 metric tons of coal or 1500 barrels of crude oil. The total volume of water in the oceans is about 1.5 billion cubic kilometers. Since one cubic kilometer is 10^9 cubic meters, we have a storehouse of energy in the ocean equivalent to 4.5×10^{20} tons of coal. As the initial supply of recoverable coal is about 7.6×10^{12} metric tons, we have about a hundred million times more energy potential in the water of our oceans. A possible deuterium-deuterium fusion reaction could provide 10^{17} kilowatt-hours of electric energy per year for 6000 years before it would deplete the deuterium content of the seas.

Tritium is not as abundant as deuterium in nature. It is radioactive and exists in tiny amounts only, if at all. To supply the proposed deuterium-tritium fusion reaction, tritium has to be generated in some way. The possibility of making tritium exists through bombarding lithium with neutrons. By this nuclear bombardment, lithium splits up into tritium and helium. Lithium is found in nature although it is considered to be a rather rare element. Some coarse-grained volcanic rocks, called pegmatites, contain it. So do the salts of salt lakes. We do not know exactly how much lithium may be recoverable on the Earth because we have not been searching extensively for it so far. Our present world resources, as far as we know, may be put to about 20 million metric tons of the ore—much of it (about 95%) is in the United States. Only part of this lithium, the so-called lithium-6, can be used for fusion. Some 7.5% of the elemental lithium is of this type. Therefore, it appears that we have something like 670,000 metric tons of lithium-6 for fusion reaction. This would be enough to generate about 2.2×10^{23} joules of energy. Presumably a worldwide search for lithium can bring about more abundant supplies. But even if only our present supply were available for the future, there is enough lithium to supply fusion energy about equivalent to the energy of the initial fossil fuel contents of the Earth.

Fusion energy would be considerably less objectionable from environmental standpoints than nuclear fission. The only radioactive element generated in the process is tritium. Tritium has a relatively short half-life of 12 years. It is an intermediate product in the process and will not end up in waste depositories as many of the fission products do. It is much weaker than, say, plutonium, and presents considerably less environmental danger. The problem with tritium is mostly technical. It destroys present-day metals; therefore, its containment and handling will require new materials and technology—a source of considerable engineering research. The use of vanadium or niobium was suggested in structural materials as both of these resist the structural degradation caused by tritium.

While theoretically we know the basic physical requirements of fusion, our contemporary technical knowledge is far from being sufficient to build a fusion reactor (98). To attain fusion, we have to generate fantastically high pressures and temperatures, exceeding those on the sun (103). The ignition temperature required for fusion is between 100 million and one billion degrees Celsius. Once the materials ignite, they must be contained for at least a small fraction of a second while the fusion occurs. During this time, enormous pressures are present in the high-temperature plasma of fusion fuels. Structural materials to withstand these effects are nonexistent at present.

There are several methods physicists think could be used for generating and containing the ignition temperature and pressure needed for fusion reactions. The two most likely contenders are the magnetic containment method and the high-power laser method (25) (72).

The magnetic containment method was first demonstrated by Soviet scientists who devised a device called tokomac. Using an electromagnetic heating method around a doughnut-shaped (toroidal) space in which the fusion reaction is to take place, the plasma of deuterium and tritium is heated to the required temperature. The magnetic field not only heats the ionized plasma to the required temperature but also contains it. If fusion is attained, the resulting free neutrons would be absorbed by a liquid lithium blanket. The lithium would be used to carry away the heat generated in the fusion, transferring this heat through a heat exchanger, forming steam. The steam would be used in a conventional steam turbine to generate electric power. The neutrons captured by the lithium blanket would change some of the lithium into tritium, which, in turn, would be separated and injected into the toroidal magnetic containment vessel along with the deuterium to provide fusion fuel. For a commercial generator, a toroidal tokomac would have to be of considerable size; a major diameter of 60 feet and a minor diameter of 24 feet is envisaged. For experimental reactors, a toroid of one-quarter of this size may be sufficient. The device must also contain superconductive magnetic coils, containment vessels for the lithium blanket, neutron shielding and reflecting material, thermal insulation, cooling ducts, and structural support. Much of this design exists only in the conceptual stage, most of it should be subject to intensive research before fusion power will be a reality. Existing experimental tokomac devices are not yet able to reach the required density of the plasma to initiate fusion. Recent breakthroughs reported by Princeton University researchers show promise that the method is a feasible way to generate fusion power (5). Their device first induces electric current into the doughnut-shaped plasma (an extremely high-temperature gas of fusion elements), heating and confining the gas in the magnetic field. The plasma at this stage is heated to 10 million degrees Celsius. Next the plasma is compressed by pulsed magnetic field. The temperature rises to 25 million degrees Celsius and the density reaches a hundred trillion particles per cubic centimeter.

In September 1974 researchers at the University of Texas attained 111 million degrees Celsius for one-millionth of a second with a density of 1000 billion particles per cubic centimeter in their "Texas Turbulent Torus." Although these temperatures and pressures are high, they are still not high enough to generate a fusion reaction. But few scientists doubt that successful demonstration will be reached in the coming decade showing capability in attaining controlled fusion (63).

Another method for the compression of fusion fuel involves powerful laser beams (107). While conventional laser beams—a high concentration of pulsating light beams—are wholly insufficient to achieve the required energy concentration, research on military application ("death rays") resulted in considerably more powerful devices. Further research on these would potentially result in lasers of sufficient power to initiate fusion. The principal advantage of the laser beam is that it itself represents a simply controllable heat source. Instead of producing plasmas in large volume and then deforming them with appropriate magnetic fields, the energy of a pulsed laser beam can be concentrated optically on a solid target—a small deuterium-tritium "ice" pellet. Lasers for this purpose must generate about 10^6 joules of energy (23). Presently available lasers, particularly the neodymium-glass laser system, can furnish only 100 joules in a short time period of one nanosecond.* This is sufficient time to attain fusion but the energy output should still be increased manyfold to be able to initiate fusion in the target pellet. Research for more powerful laser systems may result in heat energy that is sufficient to explode the pellet. The resulting blasting of the pellet would cause the internal parts of the target to be compressed to high pressures during the pulse period of the laser. The principle of internal pressure

---

*One nanosecond is one thousand-millionth of a second.

increase due to such blast is called "implosion." The heat and the implosion pressure is expected to create fusion in the fuel pellet. With sufficient financial support, researchers expect a demonstration fusion plant to be capable of operation by the year 1995 (64).

Obviously, the technical difficulties of fusion energy are enormous (113). Presently, fusion energy is at about the same level of development as nuclear fission energy utilization was 30 years ago. With reasonable research support by the federal government, scientists think that fusion will be a commercially useful technology by about the year 2000 or shortly thereafter. Up to 1974, research support was at about $30 million per year for fusion; compared to over $560 million of fission expenditures per year, this was rather small. If the 1973–74 energy crisis results in a dramatic increase in research funding, fusion research may be accelerated considerably. By the year 2000, fusion research expenditures will total about $1 billion, versus over $100 billion for nuclear fission. On this basis fusion energy technology, when developed, will find a mature fission energy technology covering the market—a tough competitor. From a practical standpoint, on the other hand, nuclear fission is a feasible commercial reality today, while fusion is only a scientific probability that may or may not be feasible for any number of yet unknown reasons. But the dream of limitless power will in all probability induce mankind to carry fusion energy development to its successful completion (36).

CHAPTER 13

Other Techniques of Energy Conversion

The bulk of our electricity is produced by conventional energy converters that are based on mechanical, indirect conversion of energy. The chemical energy of our fossil fuels is first converted into heat energy. The heat energy is then converted by turbines into mechanical energy which, in turn, produces electricity by generators. The efficiency of these systems is low—the step involving the generation of mechanical energy results in a 70% loss of energy. As a consequence of this, for many years scientists and engineers have been seeking ways to convert energy directly into electricity without the use of an intermediate mechanical energy converter.

It is interesting to note that not one of the advanced methods of direct energy conversion is really new in theory. These concepts were developed a long time ago, along with the development of classical physics. But to make these ideas work required a technological sophistication that became available only in the past few decades.

The main groups of direct energy converters are the photoelectric, thermoelectric, thermionic, magnetohydrodynamic, and electrochemical devices. Their names indicate the physical processes by which they work. The converters listed above are those by which we now can produce electric energy in quantities sufficient for practical use. Other physical effects producing electricity are piezo-electric, piroelectric, fission-electric, thermomagnetic, and chemomagnetic effects. These produce small amounts of current and are used mainly in scientific measuring devices rather than as energy producers.

Photoelectric energy converters (10, 116) transform the energy inherent in light into electricity. In 1839 A. E. Becquerel, a French scientist, found that electric current is produced if light is directed to one of the electrodes in an electrolytic solution. Forty years later a similar photovoltaic effect was recognized in solids. The studies of Germany's Max Planck in 1900 proved that light is an electromagnetic wave. Five years later, Albert Einstein realized that light is a current of photons—electromagnetic quanta of energy. They may react with the electrons of some pure metals transferring some of their energy and thereby releasing electrons from these metals. The flow of these released electrons create an electric current. Fifty years passed until this photoelectric effect was developed to the point of practical usefulness. In 1954 Bell Telephone Laboratories demonstrated the first photoelectric cell that had an acceptable conversion efficiency.

A photoelectric or solar cell may be built by splitting a chemically pure silicon crystal into two extremely thin layers. One of these layers is treated with boron, the other is treated with phosphorus, then the two layers are fused together again. The atomic bonds of phosphorus and silicon result in one excess electron. This material is called an "n"-type silicon. Conversely, the atomic bond of boron and silicon results in one less electron—an electron hole—and is called a "p"-type silicon. The two dissimilar silicon layers provide a material with an internal electric field in which electrons flow across the bond to rectify the imbalance of electron charges. If the junction of the prepared silicon layer is flooded with the photons of light, it tends to counteract the internal flow of electrons, creating a continuous flow of electric current through the junction. By placing electrodes on the "p" and "n" sides of the silicon cell, a measurable electric current can be obtained (116). The conversion efficiency of silicon solar cells depends on the thickness of the silicon layers. The thinner the cell the more efficient it is. Present-day silicon cells are about 45×10^{-3} centimeters thick. By developing thin film solar cells of 10^{-4} to 10^{-3} thickness, the efficiency of their energy conversion could be raised up to about 10%. The cost of pure silicon and the complexity of its preparation make silicon cells expensive. For this reason, their use is largely limited to space applications. Proposals for their large-scale terrestrial use are discussed in Chapter 8. Cheaper but less efficient are the cadmium sulfide solar cells.

The cadmium sulfide (CdS) cell is simple in design. A CdS cell is composed of four basic layers. A metal foil substrate forms the base electrode upon which cadmium sulfide is vacuum evaporated as a thin film. On this layer, copper sulfide (Cu_2S) is electroplated and then another metal electrode grid is placed on top. Transparent plastic foil is laminated on this cell to encapsulate it for protection. As light passes through the film sandwich, the photons activate a flow of electric current that is picked up through the metal electrodes. The efficiency of the CdS solar cell is less than that of the silicon cell but its cost is significantly less. Practical use of this system was demonstrated in solar-electric homes both in the United States and in the Soviet Union. "Solar One," a solar electric home built by the University of Delaware—shown in Fig. 13.1—operates with these cells.

Thermoelectric converters, as the name implies, convert heat energy directly into electricity. As early as 1821, T. J. Seebeck reported to the Prussian Academy of Science that if the welded junction of two wires made of dissimilar metals is heated a measurable electric current can be observed between the unheated ends of the wires. Another form of the thermoelectric effect was by a French watchmaker, J. C. A. Peltier, in

Fig. 13.1   Solar One, the first experimental solar electric home. (Courtesy of the University of Delaware.)

1834. By letting current through a welded junction of two dissimilar wires, the temperature of the junction changed. In 1838 E. Lenz of the St. Petersburg Academy demonstrated the Peltier effect by freezing a glass of water in which a bismuth-antimony wire junction was placed through which he passed electric current. Later Lord Kelvin put the Seebeck and Peltier effects on a proper theoretical foundation. He also established a third effect, now the so-called Thomson effect. This heating or cooling effect appears in a homogeneous conductor in which electric current passes in the direction of a temperature gradient. By building a series of semiconductive thermocouples to form thermopiles, repeated attempts have been made to generate power since the late 1800s. Using solar power for heating, Maria Telkes attained a 3.35% conversion efficiency in the 1930s. The efficiency of a thermoelectric generator depends on the right selection of thermoelectric materials and on the temperature difference between the two ends. The progress in material development during the past few decades made it possible to build thermoelectric converters of practical significance. In 1956 A. F. Joffe of the USSR developed a thermoelectric converter driven by a kerosene lamp (10). His research indicated that tellurides and selenides are superior thermoelectric materials. Recent research is directed toward silicium-germanium alloys that have conversion efficiencies exceeding 10%. These alloys are capable of operating at temperatures up to 1950°F.

Thermoelectric converters are under study for applications in fossil fuel electric power plants as secondary, topping generators, in space and in terrestrial power stations using nuclear heat, and in solar-powered generating units for spaceships (57).

The first major thermoelectric converter was put into use in 1964 in the USSR (19). This device, called Romaska, was made of silicon-germanium semiconductor units built into the shield of a nuclear reactor where the temperature ranged up to 1000°C. In the United States the most modern multihundred-watt thermoelectric generators are developed for space application. The 1973 power levels are about 150 watts, weighing about 90 pounds. The heat source is a Pu-238 isotope, operating at a temperature of 1300° Kelvin. The size of such a converter is 17 inches in diameter and 23 inches in height. In the future, uranium-zirconium reactors are expected to supply thermoelectric power for space applications at the range of electric power of 5 to 10 kilowatts in the 1970s and even up to 20 kilowatts in the 1980s. (In the late 1980s thermionic converters will be capable of generating 100 to 300 kilowatts of electricity in space.) Another type, the uranium-zirconium hybrid reactor consists of a cluster of cylindrical fuel elements containing uranium-zirconium hybrid rods within nickel alloy cladding tubes. The nuclear heat is carried out of the reactor by a sodium-potassium liquid metal and delivered to the thermoelectric power conversion system. The coolant operates at 1200°F. The thermoelectric modules are made of lead-telluroid material (2).

Solar-powered thermoelectric generating units were proposed as a source of electricity in satellites on a solar probe mission where silicon cells would not be efficient (10). The proposed flat plate thermoelectric generators would use silicon-germanium thermoelements. Two such panels of 23×14 inch size each would provide 150 watts of power at a voltage of 28V.

In general thermoelectric converters are limited in efficiency to about 10% to 15% (52). For terrestrial applications, such efficiencies are not competitive. Hence it is unlikely that thermoelectric generators will play a significant part in the next generation's commercial electricity production except perhaps as a topping device in conjunction with more conventional thermal conversion systems.

Thermionic converters operate on the principle that electrons may be emitted from the surface of a heated metal if the heat energy absorbed is sufficient to overcome the bonding forces of the surface atoms. This effect was first noted in the middle of the 18th century by Charles DuFay through his recognition of the fact that space in the vicinity of a red hot body is electrically conductive. In 1853 A. E. Becquerel reported the results of an experiment in which he measured a flow of electric current through a heated space between two platinum electrodes. Thomas Edison submitted a patent application in 1883 describing the generation of current in a wire passing through the vacuous space of a lamp, but he soon lost interest in the subject. A German university student, W. Schlichter, in a thesis submitted in 1915, recognized the possibility of using diodes, such as common radio tubes, as thermoelectric converters (19). Such tubes contain a heated cathode, which emits electrons through a vacuum chamber to an anode.

For the purposes of practical thermionic energy conversion, the vacuum chamber of common diodes is inefficient. But if the electrons emitted from the heated cathode surface travel through a chamber filled with either an electron gas or ionized vapor and then they are collected by a cooled collector, the flow of electrons creates an electric potential between the electrodes. The result is that if the emitter and collector are connected through an external circuit as shown in Fig. 13.2, electric current will be observed. The chamber in practical applications may be

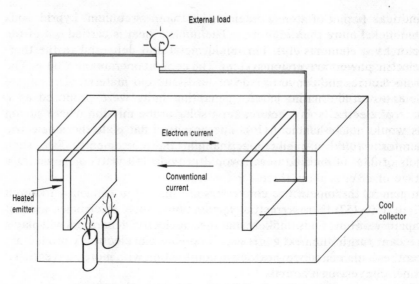

Fig. 13.2 Conceptual drawing of the thermionic converter.

filled with xenon, krypton, or neon gas. Vaporized cesium is also used for this purpose. In either case the result is better conductivity through the space (93). Recent research indicated that the best emitter electrode materials are rhenium, tantalum, and tungsten. Collector materials used include nickel, tantalum, molybdenum, stainless steel, barium oxide, and others. The temperature required for thermionic conversion is high, up to 3000°F.

The development of modern thermionic converters dates back to 1956 when G. N. Hatsopoulos described two types of such energy converters in his doctoral thesis at MIT (10).

In terrestrial applications thermionic energy converters are being demonstrated in the United States to be useful in conventional power stations for topping (19). It is known that a large portion of the heat energy is lost in a conventional power plant. The usable temperatures with conventional steam systems do not exceed 1000°F. Fossil fuels generate much more than that, making it possible to utilize this heat first in thermionic converters that work on high temperatures. The heat rejected from these converters is then channeled to the conventional system. By building thermionic converters into the system, the power station's efficiency can be improved by about 10% at a relatively low cost. On a similar basis thermionic converters were used successfully to increase the output of atomic reactors. Early results indicate that overall efficiencies of more than 45% could be attained by adding the converters to the heating elements (19). Thermionic converter-generators producing alternating current are now marketed by the General Electric Company. In these devices the converters drive a low-voltage direct current motor that rotates an alternating current generator on the same shaft.

To replace the noisy conventional engine-driven portable electric generator, a flame-heated five-kilowatt thermionic power supply was developed in 1966 by American researchers (10). Forty-eight 100-kilowatt modular diodes connected in series are used in this unit. The overall size of this converter is about three cubic feet, and its weight is 170 pounds. It consumes 3.1 gallons of gasoline per hour.

Nuclear-fueled thermionic converters were developed for various space applications. For a 125-kilowatt electric energy source, 162 modular fuel elements are required, each fuel element consisting of six diodes. In space these elements are heated by plutonium isotopes (24). There is considerable research activity to perfect these devices (2). It is generally expected (10) that the present-day 15% efficiency of thermionic conversion will be improved to about 23% within the coming decade. In this case the thermionic topping of conventional power stations will become feasible.

Magnetohydrodynamic converters change the internal energy in a hot gas into electric energy without the need of a gas turbine and an electric generator. In the MHD generators the gas itself is the conductor that passes through an external magnetic field. The electric current generated by virtue of the magnetic induction is taken off the gas by electrodes positioned at the walls of the gas duct. Therefore, there are no moving mechanical parts in the MHD generator.

The principle of operation of the MHD generator originated from M. Faraday, an English scientist. Postulating that his law of magnetic induction must be valid for electrically conductive liquids as well as solid conductors, in 1836 he tried to measure induced current between two copper plates lowered into the Thames River from a bridge. He thought that the salts dissolved in the water should make it conductive and that the Earth's magnetic field would be enough to induce a flow of electricity. His concept, of course, was correct. But his lack of knowledge about the properties of ionized fluids and about the required magnitude of the magnetic fields prevented his success (19).

The first practical MHD generator was developed in the United States by the Westinghouse Corporation in 1940. This early device was called K

& H generator after the owners of its 1936 patent, two Hungarian scientists, B. Karlovitz and D. Halasz. It worked with flue gases heated up to 1000°C. At this temperature, the gases were ionized to some degree. Ionization was also increased by electron beams. The gas was passed through a circular duct within a perpendicular magnetic field. The efficiency of this early MHD generator was limited by the lack of suitable structural materials to resist the higher temperatures that would have been necessary. The Second World War interrupted these experiments and they finally ended in 1946.

Serious studies of magnetohydrodynamic power generation were begun by A. R. Kantrowitz at Cornell University and later at Avco Everett Research Laboratory in 1956 (98). Temperatures up to 2000°C were attempted. To increase the strength of the magnetic field, superconductive magnets were proposed. An experimental generator built in 1959 produced 11.5 kilowatts of power. The device used an ionized argon gas jet as a 3000°K thermal energy source. Later, Avco Corp. built a 1500-kilowatt model using the combustion products of alcohol and oxygen. In the mid-1960s an even larger device was built with the sponsorship of the US government. This one had a capacity of 32,000 kilowatts, 8500 kW of which was used to power the magnet. In the USSR the first successfully operating MHD generator was demonstrated in 1972. It generates 25 megawatts of electricity topping a 50-megawatt conventional gas-fired turbine. The USSR is reportedly planning a giant MHD generator at a cost of $300 million. In the United States proposals are under consideration to use nuclear heat. So far, however, nuclear reactors producing the high temperatures necessary for sufficient MHD operation have not been developed.

There are two basic concepts for the arrangement of MHD devices. One operates in a simple open cycle. In this the fuel and an oxidizer are burned and a few percent of easily ionizable seed material are added. The combustion products are accelerated by a nozzle, passes into the generator and exhausted into the atmosphere as shown in Fig. 13.3. Since seeded combustion gases are not sufficiently conducting below about 4000°F at atmospheric pressure, the flame temperature must exceed 4000°F to extract any electrical power. Fuels with high flame temperatures are ethyl alcohol, kerosene, hydrogen, methane, cyanogen, and others.

Another way to operate MHD generators is the closed loop cycle. This prevents pollution of the environment. For example, the Soviet device mentioned above consists of a vapor reactor, an injector, the generator

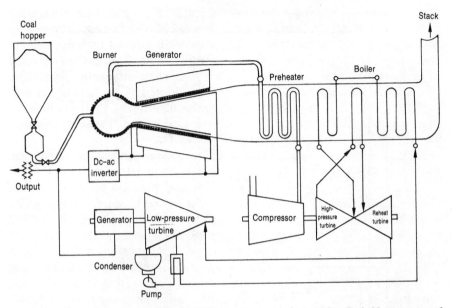

Fig. 13.3 Flow chart of a coal-fired MHD steam power plant. (After Ref. 99, courtesy of McGraw-Hill Book Co., Inc.)

itself, and a cooler. Liquid potassium is used as a working fluid. It is heated in the vapor reactor to 900° Celsius, with about 15% of the liquid being converted into vapor. The mixture of hot liquid-vapor potassium emerges from the reactor at high pressure (three times atmospheric pressure) and is allowed to expand through a nozzle to acquire a high velocity—about 1500 feet per second. The high-velocity mixture enters the injector in which a stream of cooled potassium condenses the vapor portion of the mixture, producing a flow of liquid metal at high pressure. As the potassium flow passes through the poles of the magnet in the MHD generator, electric energy is generated. The flow then returns to the vapor reactor, closing its loop.

In spite of the significant technological difficulties that are to be overcome before MHD power generation becomes a conventional method, the principle offers great advantages for the future. Topping conventional coal-fired power plants, as shown in Fig. 13.4, would increase the overall conversion efficiency by 10% to 20%. The cost of installation does not appear to be higher than that of an equivalent conventional power plant, yet the savings in fuel are significant. Because of their basic characteristics, MHD devices are most applicable for the

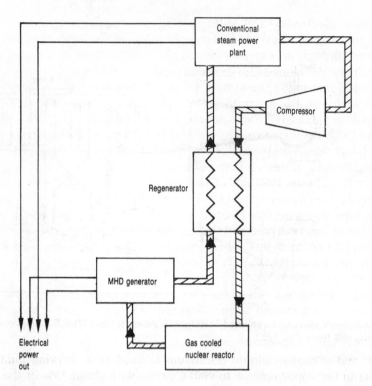

Fig. 13.4 Flow chart of a closed loop nuclear-MHD-conventional steam power plant.

production of large amounts of power. A generator producing the equivalent of ten large conventional utility plants appears wholly feasible from technical standpoints (11). MHD generators are also advantageous from pollution control standpoints. The waste heat from an MHD-topped conventional power plant rejects less heat, thus reducing thermal pollution. To recover the seed material, electronic precipitators are integral parts of its design; this reduces the fly ash in the air. Therefore, it is very likely that magnetohydrodynamic generators will receive more and more attention in the future.

Electrochemical energy converters are called fuel cells (1). Fuel cells are different from batteries. While batteries store electricity, fuel cells generate electricity from chemicals (8). The first fuel cell was built in 1801 by Humphry Davy, an English scientist (10), using carbon and nitric acid. In 1849 another Englishman, W. R. Grove, proved that electrolysis of water is a reversible process. He proceeded to build a fuel cell using two platinum electrodes and sulfuric acid as the electrolyte. Conducting hydrogen to one electrode and oxygen to the other resulted in an electric current through an external circuit. In 1860, a French scientist named Lechanche developed Grove's idea further by inventing a porous carbon electrode that made the system more efficient. The successful development of the electric generator in the second part of the 1800s resulted in a setback in aggressive fuel cell research during the next several decades. In the 1930s extensive studies were performed by F. T. Bacon of England, which laid the groundwork for modern fuel cell design. He was the first to demonstrate a series of fuel cells on the scale of several kilowatts of power (16, 21). In 1960 Allis Chalmers Corp. developed a practical fuel cell system on a commercial scale (8). One of the most common fuel cells is the type that generates electricity from hydrogen and oxygen or air. The successful use of hydrogen-oxygen fuel cells was shown in the Gemini and Apollo space flights. The power output of these units was about of one to two kilowatts. The efficiencies of these systems are in the range of 50% to 70% (2).

Lately, numerous researchers have produced dozens of different fuel cell systems for various applications (16). The main types of these may be grouped as follows: hydrogen-oxygen fuel cells (38), hydrocarbon-oxygen (air) fuel cells, regenerative fuel cells. Other types of fuel cells include metal-oxygen, ammonia-oxygen, carbon monoxide-oxygen, organic compounds-oxygen, hydrogen-halogen, and several additional types (21, 38). Of the various fuel cell types the hydrogen-oxygen fuel cell is by far the most developed one and holds the best potential for future applications. Hydrocarbon-oxygen fuel cells are thus far considerably less developed than the hydrogen-type fuel cells. The cost factor is high due to the use of platinum as the electrode. Further research is expected to solve these problems (2). These cells may make it possible to generate electricity for home use from natural gas or coal-based synthetic gas. This would eliminate the need for the low efficiency conversion of hydrocarbons first into thermal energy, then the thermal to mechanical, then the mechanical energy into electricity. Also, it eliminates electricity transfer losses. United Aircraft announced a $42 million venture in 1973 to develop a 26,000-kilowatt fuel cell as a new way to generate electricity.

Hydrogen, as a secondary energy carrier is under considerable research at present. The most commonly used hydrogen-oxygen fuel cell has many advantages as a secondary electric energy producer (8). Producing hydrogen by the electrolysis of water is a convenient choice for intermittent energy producers such as tides, winds, or solar heat. Nuclear reactors would be able to produce hydrogen in great quantities. Storing hydrogen in a compressed form, as a gas or liquid, perhaps at low temperatures, is a well-established technology. Even underground storage is said to be feasible in exhausted gas or oil reservoirs. Hydrogen can be delivered to the user through pipe in the same manner as our existing gas transport pipe network. There it could be directly burned to provide heat or it could be used in a fuel cell to produce electricity. Hydrogen is nonpolluting and the only by-product of a hydrogen-oxygen fuel cell is water. Many scientists believe that in a few decades, when fossil fuel supplies are badly depleted, the nation will run almost exclusively on hydrogen. Presently, automobiles are receiving the most attention from researchers. Almost any type of engine can be run on hydrogen. At the Los Alamos Scientific Laboratory and at the Jet Propulsion Laboratory at Pasadena, California, as well as at the Brookhaven National Laboratory on Long Island, research is underway to use liquid hydrogen in place of gasoline. Although such studies date back to about 1930, the major shortcoming of the hydrogen-driven car is still the bulky hydrogen storage required. To match the range of a 200-gallon gasoline tank with liquefied hydrogen, about 3000 pounds of hydrogen tanks would be required. To overcome this difficulty, scientists are looking at a metal-hydride system. This is based on the fact that pure hydrogen gas can be adsorbed in a condensed layer on the surface of certain metals such as iron, titanium, magnesium, or aluminum. A hydrogen-charged metal "sponge" could hold a significant supply of hydrogen. To refuel such a car, a driver could have his expended metal-hydride system replaced at a gas station. The engine needs only small carburetor adjustments to run on hydrogen. This, being a clean fuel, would cause less degradation to the engine than gasoline does and would eliminate air pollution.

Hydrogen has many other possible uses in energy supply. It may be used as industrial fuel or for fuel in local conventional power stations. It is a necessity in the manufacturing of synthetic gas from coal. There is a considerable use of hydrogen in the chemical industry. The conversion efficiency of a hydrogen fuel cell is about 70%, twice that of the conventional electric generator turbine's best efficiency. It appears that hydrogen can well replace all fossil fuels in practical applications. Methods for producing hydrogen include electrolysis from water, sunlight, and biological methods, heat and enzymes, chemical processes, etc. Processes for hydrogen and oxygen production that require only thermal energy are under development in Europe. These processes involve a closed set of chemical reactions. With only heat and water input, hydrogen and oxygen are produced without any reaction products and wastes, all of these are recycled in the process.

Methanol or wood alcohol is another secondary energy carrier that may have a bright future. It is a clean-burning fluid that can be produced in a variety of ways from many sources including trees and other vegetation, garbage, coal, and other hydrocarbons. It is easy to handle and can be stored and transported in tanks and pipelines without the need for new technology. It may be burned directly—the well-known Sterno fluid is methanol. It was in wide use as lighting fluid before petroleum came on the market. The street lights of Paris were fueled with wood alcohol during the middle of the 19th century. During both World Wars, methanol was resurrected in Europe. Thousands of cars were converted to run on methanol. When mixed with gasoline up to 15%, methanol fuels automobile engines without conversion. It has a higher octane than gasoline but is a nonpolluting fuel. Chemically, methanol is a carbon monoxide molecule bound with two hydrogen molecules. It therefore may be looked upon as a convenient liquid carrier of hydrogen energy.

As the foregoing review of novel techniques in energy production shows, one could expect to see a great variety of new processes and unconventional devices in the coming decades. The research activity in this field is very active and new developments are announced practically every week, making yesterday's expectations obsolete by tomorrow. The developing new technology improves the efficiency of power generation on the one hand. On the other hand, it tends to change the pattern of energy utilization by which a slow but relentless change is caused in the lifestyle of the people.

CHAPTER 14

The Challenge

The 18th-century English poet William Blake said in "The Marriage of Heaven and Hell" that, "Energy is the only life ... Energy is Eternal Delight." This is not an overstatement. Energy is of overwhelming importance in the quality of our life. Fortunately, as the previous chapters demonstrated, there is no overall shortage of energy resources. The energy crisis that we are talking about in these early years of the 1970s is actually an energy allocation and utilization crisis. We could more than double our energy utilization without increasing the raw energy production by an improvement of the efficiencies of the multitude of energy conversion devices and energy extraction technologies. By developing new types of converters, we could change our energy picture in a drastic way. In the process we may also have to change our lifestyle. Steps toward this should be taken soon. Our country as well as the rest of the industrial world grows at an increasing rate. In our technological age more and more people attain the increased benefits of progress. As one expert quipped in 1973, "There is no shortage of gasoline, there is only an overabundance of automobiles."

The rest of the world is growing at an even faster rate than the United States. Even with the faster rate of growth, economists estimate that the world's per capita energy use in the year 2000 will be only one-quarter that of the United States. This discrepancy will put considerable stress on political conditions in the world (35). Today the economy of some parts of the world outside of the United States (the best example being Western Europe), is growing at a high rate, while the growth in underdeveloped countries such as India is almost imperceptibly slow. In order to bring the

145

rest of the world to a level comparable to the United States, the energy production should be increased tenfold on a worldwide basis. Energy needs industry to use it, capital for these developments is scarce, raw materials are even scarcer. Markets for these products would be virtually nonexistent unless some major revolutionary changes occur in the underdeveloped countries. One of the major problems of such universal development would be the tremendous strain it would put on the world's existing fossil fuel resources. Most of the advanced technological countries, with the exception of the Soviet Union and Canada, are not self-sufficient in energy (76). The table below shows the percentages of local availability of oil and the total energy resources in major countries in 1971 (76):

Country	Oil	Total Energy
United States	74%	89%
Canada	98%	110%
Japan	0%	11%
United Kingdom	2%	53%
Germany	7%	51%
France	5%	22%
Italy	6%	15%

The supplies are certainly great around the world to fulfill these needs. Yet we know that these supplies are finite. Most of the world's oil production, for example, is shared by the United States, Europe, and Japan. It would surely be insufficient to provide the gasoline for all other nations if they would at least approach the degree of development of these leading regions. A major problem, therefore, is that much of the world's energy supply is still based on fossil fuels. Fossil fuels, as the previous chapters showed, are limited and the supply is fast running out (55). They are considered as capital, their use is like living on past savings. Geothermal and solar energy (also, to an extent, fusion power) exist in much greater quantity. These (the first two, at least) could be considered interest, and using them is like living on profits of investment (87).

From this, it follows that the future of our energy supply must be based on sources other than fossil fuels—coal, oil, and gas—or even nuclear fission of uranium (124, 125).

Lately some theoreticians (88) have been talking about the concept of the "nongrowth" society, claiming that a society based on growth has only a limited life. In fact, there are three alternatives for society—one based on "future of expanding power," another on "future of constant power," and the third based on "future of receding power." Expanding power may be limited in its future. It will result in tremendous international stresses. Unlimited growth will lead to competitive pressures between countries that may lead to wars. Total wars between advanced countries would lead to total destruction. Only through powerful control by international agencies could the proper economic planning, coordination, development, and the orderly growth of civilization be assured (51).

After an initial growth, the world will most likely level off its energy utilization. From a state of steady expansion, energy utilization will turn to a steady climax. The advantages under those conditions will be with those who can do the most with the least, who can do best with what they have. As fossil fuels run out, those countries and regions that have developed the best technology for utilizing the constant flow of "income" energy will have the advantage. The challenge consists in developing the sufficient technology to extract energy from nature without ecological dangers, such as air, water, thermal, or radioactive pollution. If we can do that while our fossil fuels last, we assure our future. If we don't, we can look forward to a phase of receding power that would bring about a scenario that we would not presently care to discuss.

There are some fundamental problems in American society that prevent us from making the necessary changes in the energy utilization and allocation pattern. None of these problems are unsolvable. To solve them, however, they must be brought into focus.

First, we must recognize that meeting future needs in a realistic manner needs far more overall planning and coordination than is exercised by our governmental agencies at present. While our agricultural production is regulated by a monumental federal department, control of energy production and utilization (until the recent establishment of the Federal Energy Administration) was fragmented into dozens of organizations. Each one of these appeared to be operating as a special interest group lobbying for the industries related to and "controlled" by the individual agency. The discordant interests seem to be counterproductive from the standpoint of the strongly correlated energy policy sorely needed by our country. In contrast, foreign countries are more advanced from this standpoint. Not only does Soviet research and development appear to surpass our progress but major countries of the free world grow at a rate that exceeds our growth strictly because of a more orchestrated utilization of their technological resources. In the United States the energy research, development, and allocation patterns still seem to be controlled by a multitude of competing industrial firms, in the best tradition of 19th-

century capitalism. The traditional concepts of profit, interest of investment, and other capitalistic ideas are in full force in the American energy
scene. Meanwhile, in other phases of our life these concepts were long
surpassed. In our federal "deficit spending," our agricultural controls and
subsidies, and in a broad spectrum of the rest of our economic life, the
governing ideas appear to be quite different. We still, sometimes, give
lipservice to the old capitalistic concepts if we try to make progress in
some fields. The ridiculous artificiality of the "cost–benefit" studies of
water resources projects of the US Army Corps of Engineers serves as a
good example of these. How could we devise a way to promote our
general interests in areas such as geothermal energy utilization, hydrogen
fuel cell utilization, coal gasification, and other important areas when we
still do not have a major industrial strength to "lobby" for these interests?

The second problem we face is the poor showing of our country in
developing our intellectual resources that would assure our progress in
the future. The USSR is producing three times as many engineers as we
do each year. Their engineers are at least as well trained as ours, or better.
The social status of engineers is significantly higher in all parts of the
world than in the United States. In all other countries, engineers are
considered at least equal to or of higher status than lawyers or doctors.
Here, engineers are rated well below these, even below architects and
stockbrokers. True professional status is not accorded to engineers in
America, although the term professional that is freely used for hairdres-
sers and teachers covers engineers also. In reality, American engineers
are in bondage to industry, where in case of striking labor they are
considered, fraudulently, to be part of "management" by the embattled
firm. The talents of our engineers are wasted by being employed well
below the capacity for which they were educated. The average American
engineer works at an intellectual level barely beyond his sophomore
training. If engineers received more recognition and were better trained,
better paid, and free to move with portable pension systems, they could
be utilized in a much more efficient manner. This would require a greater
level of support in the form of trained technicians and draftsmen, a breed
so scarce in America.

Another waste of intellectual capacity is the lack of women in the
engineering fields. In most cilivized countries women engineers make up
almost 50% of the total. In the United States their participation is well
below 1%. What a waste of intellect! The woman engineer in this country
is headline news. In this sense she is considered to be an oddity.
Newspapers play up how she was grudgingly accepted by the workers and

how well she is doing on the job—presumably despite her sex. In reality for most engineering jobs of today, women are much more capable than men. A woman's characteristic attention to detail goes well with computers and her patience is better suited for many of the engineering routines. There is nothing manly in pushing buttons on a testing machine or seeing if a design satisfies the related standards. Indeed, it is hard to conceive of why women should not flock to the engineering profession. They could well raise our intellectual output by an amount necessary to meet the future needs of our country in the forthcoming technological age of unimaginable complexity.

Problems to Ponder

1.1 Using United States' per capita energy use for the year 1970 as a base, assume that India's population uses 5% of that amount. Calculate the amount of additional yearly energy production required in India in order to quadruple the per capita energy utilization.

1.2 From local newspaper files, research contemporary information about the 1923 oil crisis in the United States.

1.3 Using United Nations' statistical data, determine the ratio of the gross national product (GNP) and the energy consumption on a per capita basis in Nigeria, Portugal, and Iran.

1.4 From United States' census data for the past 30 years, determine the doubling period of the population growth of Florida and Connecticut.

2.1 Describe the various forms of energy present in a bouncing tennis ball at its various positions during one cycle.

2.2 In 1970 the per capita energy consumption in the United States was stated to be 230,000 kilocalories. Express this amount in five different energy units.

2.3 The Grand Coulee Dam's power station has a capacity of 1974 megawatts. How much oil, natural gas, or coal does this power generating capacity correspond to in one year?

2.4 On the basis of the Second Law of Thermodynamics, discuss the operational efficiency of an air-conditioner.

3.1 Assuming that by the year 1980 there will be 85 million wired homes in the United States, determine the total residential electric energy demand for that year based on your family's electric bill for the past month. How much growth would this represent with respect to the year 1970?

3.2 How much energy would be saved if 30 incandescent light bulbs, 100 watts

151

each, would be replaced by fluorescent tubes supplying the same output in lumens of light?

3.3 On the basis of your yearly electric, gas, fuel oil, and gasoline bills, and by obtaining all corresponding utility rates, determine the total amount of energy your family has used during the past year.

3.4 By studying the pattern of your family's yearly energy use, list various possible measures for potential savings in case energy costs increase by a rate of twofold and/or fivefold.

3.5 Assume that there are 100,000 automobiles in use in the United States for private transportation, consuming gasoline on a basis of 15,000 miles per year and 15 miles to a gallon. Determine the daily rate of gasoline usage. Assuming that on the average there are two persons in the car and that 10% of the trips could be shifted to bus transportation, determine the possible savings in passenger-miles and in gallons of gasoline per day.

4.1 Assuming that anthracite coal costs $20 per ton, compute the cost of its heat energy in dollars per BTU based on anthracite's calorific value.

4.2 If the available natural gas reserves in the United States would be exhausted during a period of 13 years and on the ground that we presently use about 30×10^{12} cubic feet of gas a year, calculate the necessary yearly expenditures on coal gasification plants and the required increase in coal production in order to maintain our present gas supplies by synthetic gas.

4.3 List all necessary legislative and economic steps you would propose to undertake in order to increase our coal production by 50%.

4.4 Research the problem of economic losses due to air pollution and compare them to the costs of installing pollution control equipment at power stations.

5.1 Devise economic incentives for international oil companies in order to induce them to build more new refineries in the US.

5.2 Consider the various ways by which the government is controlling gasoline prices and availabilities in the country.

5.3 List reasons for or against changing the depletion allowance in the taxation of oil producers.

5.4 What possible ways could the United States use to eliminate the need for importing foreign oil?

6.1 Assuming that all American natural gas reserves should be held for residential and commercial use, devise legislative means by which other uses of gas might be discouraged.

6.2 Determine the per capita use of natural gas in the United States. Assuming a similar rate of use in Europe, determine the yearly natural gas needs of the European Economic Community.

6.3 Obtain data on current natural gas, oil, and coal prices delivered to your community and compute their cost per BTU of heating value.

6.4 By plotting gas reserves versus production data given in Chapter 6, determine the probable date by which our reserves will be reduced to a mere five years of supply.

7.1 If the average specific heat (heat energy required to raise or lower temperature by one degree) of a rock is 0.75 kilocalorie per kilogram, and its density is 2.5 grams per cubic centimeter, how much energy could be released from a 1000-foot-thick 1-square-mile-large dry geothermal reservoir if its average temperature could be reduced by 5° Celsius?

7.2 Consider the effects of possible new legislation extending the depletion allowance concept to geothermal heat.

7.3 What are the factors that prevent the rapid development of the utilization of dry geothermal reservoirs?

7.4 What possible environmental effects should be considered in the use of geothermal power?

8.1 How much solar energy falls on a square mile area in the state of Oregon during an average winter day. Using a conversion efficiency of 2.5% how much solar energy could be developed from such a solar field in terms of the daily energy need of each resident of the state?

8.2 Compute the magnitude of the average daily insolation over the land of the United States if air pollution constituents would increase the amount of sunshine reflected back from the atmosphere into space from 30% to 40%. Compare your results to data in Fig. 8.1 and draw conclusions concerning changes in our weather.

8.3 Determine the cost of yearly hot water use in your home, including the installation cost of a hot water heater. How much should the cost of a solar water heater be in order that it may be competitive with conventional systems?

8.4 List the various possible economic incentives that may spur the development of solar energy utilization. Identify those industries that may be hurt by such steps.

9.1 Discuss the reasons why hydraulic energy utilization is retarded in the United States.

9.2 By consulting topographic maps and local evaporation rates, study the technical feasibility of a depression power plant at Death Valley, California. How much power could such scheme provide and what would the cost of installation involve?

9.3 Outline technical schemes for the production of liquid hydrogen by a depression power plant at Lake Assale, Ethiopia, to be shipped by special tankers to the US.

9.4 List the physical requirements for the development of a financially feasible tidal power plant.

10.1   Assume that your are the US agent of a foreign manufacturer of wind generators. How would you organize your advertising campaign to promote sales?
10.2   Suppose you establish a wind generator company. In what sizes (capacities) would you prefer to manufacture your units and why?
10.3   By consulting local weather data from nearby airports, determine the magnitude, duration, and frequency of winds at your locality.
10.4   By contacting sellers of wind generators, determine the cost of supplying 20% of your home's electric energy requirement on a "when available" basis without battery storage capacity.

11.1   Obtain information on the licensing procedures of nuclear reactor installations. What legislative means could be developed in order to minimize delays?
11.2   What economic incentives would you use to improve the competitive position of nuclear power over conventional power plants?
11.3   By reviewing actual case histories, list the most prominent environmental objections to nuclear power plants. What legal means do environmentalists use to delay and prevent the construction of nuclear power plants?
11.4   Review the various ways proposed for nuclear waste disposal and comment on their possible environmental consequences.

12.1   By studying pertinent literature concerning the past rate of improvement in attained pressures and temperatures for fusion reaction, estimate the time by which fusion power could be technically feasible if fusion research is maintained at its current rate.
12.2   What are the reasons for claiming that fusion power is less objectionable from environmental standpoints than nuclear fission?
12.3   In what fields of technology should major progress be made before fusion power can be made a reality?
12.4   Estimate the world's energy needs by the year 2000 and the potential share of fusion power in total energy production after that time.

13.1   List the various potential applications of hydrogen as a fuel and design a flow diagram showing the production, transportation, storage, and utilization schemes required for a hydrogen-run economy.
13.2   What are the direct energy converters most suitable for topping conventional power stations? What amount of additional energy could be produced in the United States if all existing utilities were induced to utilize these techniques?
13.3   Direct energy converters were conceptually developed back in the 19th century. Assuming that rotary-type electric generators were not invented, in what possible ways would we be producing electric power? How would this have affected our present lifestyle?
13.4   Compare the schemes of a national economy run on hydrogen or methanol technology. Which would be cheaper to develop under our present technological conditions?

14.1 It has been demonstrated that there is no shortage in energy carriers in the world but that our present system of energy production and utilization cannot possibly keep up with future demands. What broad plans would you recommend for reordering our economic and technological efforts to provide ample power for the future?

14.2 Looking at the energy supply problem from the standpoint of the leader of a small, developing country having limited resources, what political and economic systems would you consider in order to best foster your country's development?

14.3 In view of the fact that the United States is in economic competition with the rest of the world and that our supply of raw materials largely depends on foreign sources, propose major national programs designed to maintain and improve our competitive edge and assure our national security for the future.

14.4 How would you, as a major policy maker of the United Nations, view the world's energy future and in what respects would you disagree with American national interests?

References

1. Adams, D. R. *et al. Fuel Cells, Power for the Future.* Boston, Mass.: Purnell, 1960.
2. A.E.C. Authorizing Legislation, Fiscal Year 1973, Hearings Before the Joint Committee on Atomic Energy. *Congress of The United States,* Jan. 26, Feb. 3, and Feb. 17, 1973, (p. 116, 164). Washington, D.C.: US Government Printing Office, 1972.
3. A.E.C. Authorizing Legislation, Fiscal Year 1973, Hearings Before the Joint Committee on Atomic Energy. *Congress of The United States,* Feb. 22, 23, 1972. Washington, D.C.: US Government Printing Office, 1972.
4. A.E.C. Cites Contractor in Leakage. *The Akron Beacon Journal,* Aug. 5, 1973.
5. A.E.C's Fusion Lab at Princeton Reports Progress in the Development of Fusion as Power Source, News Release Dec. 1, 1972. *U.S. Atomic Energy Commission,* Vol. 3, No. 49, Dec. 6, 1972.
6. Agnew, W. G. *Automotive Power Plant Research.* Presented to the Senate Subcommittees on Environment and Science, Technology and Commerce, Washington D.C., May 14, 1973.
7. Alfven, H. *Cosmical Electrodynamics.* London: Oxford University Press, 1950.
8. Alles, J. J. *et al. Fuel Cell Systems* (Advances in Chemistry Series 47). Washington D.C.: American Chemical Society, 1965.
9. American Power Conference. *Nuclear Engineering Intern,* 1971, 7, 581., 1971.
10. Angrist, S. W. *Direct Energy Conversion,* 2nd edition. Boston, Mass.: Allyn & Bacon, 1971.
11. Bartlett, D. L., and Steele, J. B. *Oil, the Created Crisis.* Knight Newspapers, series from *Philadelphia Enquirer, The Akron Beacon Journal,* Nov. 18, 19, 20, Dec. 2, 9, 16, 1973.
12. Bassler, F. Solar Depression Power Plant of Qattara in Egypt. *Solar Energy,* Vol. 14, Pergamon Press, 1972, pp. 21–28.
13. Beall, S. E., Jr. Total Energy, a Key to Conservation. *Consulting Engineer,* Vol. 40, No. 3, March 1973.
14. Beiser, A. *The Earth.* New York: Life Nature Library, Time-Life Books, 1963.
15. Bennett, I. Monthly Maps of Mean Daily Insolation for the United States. *Solar Energy,* Vol. 9, No. 3, 1965.

16. Berger, C. *Handbook of Fuel Cell Technology.* Englewood Cliffs, N.J.: Prentice-Hall, 1968.
17. Bertin, L. *Larousse Encyclopedia of The Earth.* New York: Prometheus Press, 1961.
18. Bishop, A. L. *Project Sherwood, The U.S. Program in Controlled Fusion.* Reading, Mass.: Addison-Wesley, 1958.
19. Bito, J., and Sinka, J. *Energy, Key to Our Future* (in Hungarian). Budapest: Kossuth, 1973.
20. Blackwood, O., and Kelly, W. *General Physics,* 2nd edition. New York: Wiley, 1955.
21. Bockris, J. O'M., and Srinivasan, S. *Fuel Cells, Their Electrochemistry.* New York: McGraw-Hill, 1969.
22. Boucher, C. T. G. Harnessing the Forces of Nature, Binding the Wind. *Consulting Engineer,* Vol. 41, No. 6, Dec. 1973.
23. Boyer, K. Laser-initiated Fusion—Key Experiments Looming. *Astronautics and Aeronautics,* Jan. 1973.
24. Carliss, W. R., and Harvey, D. L. *Radioisotope Power Generation.* Englewood Cliffs, N.J.: Prentice-Hall, 1964.
25. Carruthers, R., Davenport, P. A., and Mitchell, J.T.D. The economic generation of power from thermonuclear fusion. Culham Laboratory, Great Britain CIM-R-85, 1967.
26. Coal May Burn Again in Germany. *The Akron Beacon Journal,* Dec. 9, 1973, p. A7.
27. Coal to the Rescue This Winter? *The Wall Street Journal,* Nov. 12, 1973, p. 26.
28. Combustion Engineering, Inc. *1972 Annual Report to Investors.* Stamford, Conn., 1973.
29. Conservative and Efficient Use of Energy. *Hearings before the Subcommittee of the Committee on Government Operations,* U.S. House of Representatives, May 1 and 2, 1973. Washington, D.C.: US Government Printing Office, 1973.
30. Consumer's Guide to the Efficient Energy Use in the Home. *American Petroleum Institute,* 1973.
31. Cook, E. The Flow of Energy in an Industrial Society. *Scientific American,* Vol. 225, No. 3, Sept. 1971.
32. Coombe, R. A. (ed.) *Magneto-hydrodynamic Generation of Electrical Power.* London: Chapman and Hall, 1964.
33. Crisis of Converting Fuel into Power. *Business Week,* Jan. 8, 1972, p. 58.
34. Curtis, R., and Hogan, E. The Myth of the Peaceful Atom. *Natural History,* Vol. 78, No. 3, March 1969.
35. Darmstadter, J. Energy's Impact on Foreign Policy. *Consulting Engineer,* Vol. 40, No. 3, March 1973.
36. David, E. E., Jr. Energy, a Strategy of Diversity. *Technology Review,* M.I.T., June 1973.
37. Diekomp, H. M. *Nuclear Space Power Plants.* Atomics International, 1967.
38. Douglas, D. L. *Fuel Cells.* New York: Reinhold, 1960.
39. Dungey, J. *Cosmic Electrodynamics.* New York: Cambridge University Press, 1958.
40. Dye, L. U.S. Sitting on AEC Contamination Time Bomb, Scientists Say. *The Akron Beacon Journal,* July 5, 1973.
41. Dyson, F. J. Energy in the Universe. *Scientific American.* Vol. 225, No. 3, Sept. 1971.
42. Eardley, A. J. *General College Geology.* New York: Harper and Row, 1965.
43. Energy and Public Policy, Conference Board on America's Future Supply of Energy. *The Conference Board Record,* July, 1972.
44. Energy Crisis, Time for Action. *TIME,* May 7, 1973.
45. Energy Crunch: Real or Phony? *TIME,* Jan. 21, 1974.

46. Energy, Potential of Conservation. *Technology Review*, M.I.T., May 1973, p. 47.
47. Enrichment, Europe Assesses Needs, *Nuclear News*, Vol. 14, No. 8, Aug. 1971, p. 28.
48. Federal Study Stresses Energy Conservation. *Nuclear News*, Vol. 15, No. 11, Nov. 1972.
49. Fink, D. J. Monitoring Earth's Resources from Space. *Technology Review*, M.I.T., June 1973.
50. Forum on the Energy Crisis—Public Utilities. *Akron Technical Journal*, June 1973, p. 4.
51. Freeman, S. D. Outlook for the Future. *Consulting Engineer*, Vol. 40, No. 3, March 1973.
52. Gadoff, E. F., and Miller, E. *Thermoelectric Materials and Devices*. New York: Reinhold, 1960.
53. Gambs, G. C. Industry Faces New Challenge. *Consulting Engineer*, Vol. 40, No. 3, March 1973.
54. Gaucher, L. P. Energy Requirements of the Future. *Solar Energy*, Vol. 14, Pergamon Press, 1972, pp. 5–10.
55. Gaucher, L. P. Energy Sources of the Future for the U.S. *Solar Energy*, Vol. 9, No. 3, 1963.
56. Glasstone, S., and Loveberg, R. H. *Controlled Thermonuclear Reactions*. New York: van Nostrand, 1960.
57. Goldsmid, H. J. *Applications of Thermoelectricity*. London: Methuen, 1960.
58. Gwynne, P. Nuclear Relief for Natural Gas. *Technology Review*, M.I.T., July–Aug. 1973.
59. History and a Forecast of Nuclear Development in Argentina, Canada, Fed. Rep. of Germany, France, India, United Kingdom, United States of America. *Bulletin of the International Atomic Energy Agency*, Vol. 14, No. 6, 1972.
60. Hogerton, J. D. Notes on Nuclear Power. *Atomic Industrial Forum*, 2nd edition. 1970.
61. How Practical is Solar Power? *U.S. News and World Report*, Dec. 24, 1973.
62. Hubbert, M. K. The Energy Resources of the Earth. *Scientific American*, Vol. 225, No. 3, Sept. 1971.
63. Hurwitz, H., Jr. *Comments on the Prospects of Fusion Power*. General Physics Lab., General Electric Co., July 1971.
64. J.C.A.E. Assesses Effect of Added Funding on Fusion Timetable. *Nuclear Industry*, Dec. 1971, p. 45.
65. James, E. O. *The Ancient Gods*. New York: Putnam & Sons, 1960.
66. Jenkins, L., and Bandur, B. Engineers Seek to Stretch Resources. *Consulting Engineer*, Vol. 40, No. 3, March 1973.
67. Judge, A. W. *Modern Gas Turbines*. London: Chapman and Hall, 1947.
68. Kaufman, A. Managing Existing Resources. *Consulting Engineer*, Vol. 40, No. 3, March 1973.
69. Kettani, M. A., and Gonsalves, L. M. Heliohydroelectric (HHE) Power Generation. *Solar Energy*, Vol. 14, Pergamon Press, 1972, pp. 29–39.
70. Kruger, P., and Otte, Carel. *Geothermal Energy, Resources, Production, Stimulation*. Stanford, Calif.: Stanford University Press, 1973.
71. Lapp, R. E. Nuclear Energy: How Soon? How Safe? *Consulting Engineer*, Vol. 40, No. 3, March 1973.

72. Laser Implosion: Will It Speed Fusion Time-table? *Technology Forecasts*, Vol. 4, No. 9, Sept. 1972.

73. Lof, G. O. G., Close, D. J., and Duffie, J. A. A Philosphy for Solar Energy Development. *Solar Energy*, Vol. 12, No. 2, 1968, pp. 243–250.

74. Lof, G. O. G., Duffie, J. A., and Smith C. O. World Distribution of Solar Radiation. *Solar Energy*, Vol. 10, No. 1, 1966.

75. Luten, D. B. The Economic Geography of Energy. *Scientific American*, Vol. 225, No. 3, Sept. 1971.

76. McCracken, P. W. The February Agenda: A Slim Hope. *The Wall Street Journal*, Jan. 23, 1974.

77. McLean, J. G., and Davis, W. B. Guide to National Petroleum Council. *Report on U.S. Energy Outlook*. National Petroleum Council, Dec. 11, 1972.

78. Morcos, A. The Role of Probability in Nuclear Plant Design. *Consulting Engineer*, Vol. 41, No. 6, Dec. 1973.

79. Morley, F., and Kennedy, J. M. Fusion Reactor and Environmental Safety. *Nuclear Fusion Reactor Conference*, Culham, Great Britain, Sept. 1969.

80. Morrow, W. F., Jr. Solar Energy, Its Time is Near. *Technology Review*, M.I.T., Vol. 76, No. 2, Dec. 1973.

81. Morton, R. C. B. Resources: Are We Running Short? *Consulting Engineer*, Vol. 40, No. 3, March 1973.

82. Mosonyi, E. *Water Power Development*, Budapest: Publishing House of the Hungarian Academy of Sciences, 1963.

83. Natural Gas Supply: The Problem—The Solution. *Consolidated Natural Gas Company*, February 1, 1973.

84. Nephew, E. A. The Challenge and Promise of Coal. *Technology Review*, M.I.T., Vol. 76, No. 2, Dec. 1973.

85. New Action Needed to Remedy Energy Gap. *Automotive Information*, Motor Vehicle Manufactures Association of the U.S., Vol. 1, No. 5, April 1973.

86. New Look at Solar Energy. *Technology Forecasts*, Sept. 1972, p. 13.

87. *New Sources of Energy and Economic Development—Solar Energy, Tidal Energy, Wind, Geothermic Energy, Thermal Energy of the Seas*. New York: United Nations Department of Economic and Social Affairs, 1959.

88. Odum, H. T. A Print-out of the Future Systems of Man. *Natural History*, Vol. 80, No. 5, May 1971.

89. Oil Scarcity: Illusion Made in the U.S.A. *Technology Review*, M.I.T. June 1973.

90. Out of the Hole with Coal. *TIME* Jan. 28, 1974, p. 32.

91. The (Possible) Blessings of Doing Without: Time Essay. *TIME*, Dec. 3, 1973.

92. Potential of Geothermal Power. *Business Week*, March 17, 1973, P. 74.

93. Psaroutakis, J. *Thermionic Power Generation*. London: Academic Press, 1965.

94. Question of Radioactive Waste Danger Spawns Controversy. *The Akron Beacon Journal*, Sept. 2, 1973.

95. Radioactive Water in Colorado Town. *The Akron Beacon Journal*, Sept. 30, 1973.

96. Ralph, E. L. Large Scale Solar Electric Power Generation. *Solar Energy*, Vol. 14, Pergamon Press, 1972, pp. 11–20.

97. Rice, R. A. How to Reach that North Slope Oil, Some Alternatives and their Economics. *Technology Review*, M.I.T., June 1973.

98. Rosa, R. J. *Magnetohydrodynamic Energy Conversion*. New York: McGraw-Hill, 1968.

99. Rose, D. J. Fusion and Other Long-Range Strategies. *Consulting Engineer*, Vol. 40, No. 3, March 1973.
100. Saif-Ul-Rehman, M. Solar Energy Utilization in Developing Countries. *Solar Energy*, Vol. 11, No. 2, 1967.
101. Schreck, A. E. (ed) *Minerals Yearbook*. Washington D.C.: U.S. Bureau of Mines, 1970.
102. Seifert, W. W. *et al. Energy and Development, A Case Study*, Massachusetts Institute of Technology Report N. 25. M.I.T. Press, 1973.
103. Simon, A. L. *An Introduction to Thermonuclear Research*. Oxford: Pergamon Press, 1960.
104. Snyder, N. W. *Energy Conversion for Space Power*. New York: Academic Press, 1971.
105. Societie Hydrotech. de France. Six ans d'exploitation de l'usine maremotrice de la Rance. *La Houille Blanche*, Nos. 2-3, 1973.
106. Solar Energy Cited as Vast Resource. *Salt Lake Tribune*, March 7, 1973.
107. Solon, L. R. Lasers and Fusion. *Industrial Research*, Nov. 1971, p. 76.
108. Spahr, C. E. Energy Crisis. *The Ohio Contractor*, July 1973.
109. Special Fuel Conference Report, Nuclear Industry. *Atomic Industrial Forum*, Vol. 18, No. 2, Feb. 1971.
110. Sporn, P., and Kantrowitz, A. Magnetohydrodynamics—Future Power Process? *Power*, Vol. 103, No. 11, 1959, pp. 62–65.
111. Starr, C. Energy and Power. *Scientific American*, Vol. 225, No. 3, Sept. 1971.
112. Statement of Policy, Energy. *American Petroleum Institute*, Nov. 1972.
113. Steg, L., and Sutton, G. W. The Prospects of MHD Power Generation. *Astronautics*, Aug. 1960.
114. Summers, C. M. The Conversion of Energy. *Scientific American*, Vol. 225, No. 3, Sept. 1971.
115. Sun is an Energy Source We Need to Learn to Use—Editorial. *The Akron Beacon Journal*, July 3, 1973.
116. Sutton, G. W. *Direct Energy Conversion*. New York. McGraw-Hill, 1966.
117. That Other Shortage (U.S. Natural Gas). *TIME*, Jan. 7, 1974, p. 40.
118. There is Plenty of Coal—What's Behind the Holdup? *U.S. News and World Report*, Dec. 24, 1973, p. 59.
119. U.S. Department of the Interior, *United States Energy Fact Sheets, 1971*, Washington D.C.: U.S. Government Printing Office, Feb. 1973.
120. Vast New El Dorado in the Arctic. *TIME*, April 9, 1973, p. 30.
121. Vendl, A. *Geologia* (in Hungarian). Budapest: University Publishing Company, 1951.
122. Wallace, G. D. F.T.C. Blames Fuel Shortage on Government Regulations. Associated Press, Washington, June 1973.
123. Weihe, H. Fresh Water from Sea Water; Distilling by Solar Energy. *Solar Energy*, Vol. 13, Pergamon Press, 1972, pp. 439–444.
124. Weinberg, A. M. Long Range Approaches for Resolving the Energy Crisis. *Mechanical Engineering*, June 1973.
125. What Comes After the Petroleum Orgy? *Technology Review*, M.I.T., July-Aug. 1973.
126. What's Holding Up Nuclear Power? (interview with Dr. Dixy Lee Ray). *U.S. News and World Report*, Nov. 26, 1973, p. 23.
127. White, D. C. The Energy—Environment—Economic Triangle. *Technology Review*, M.I.T., Vol. 76, No. 2, December 1973.

128. Wolf, M. Solar Energy; An Endless Supply. *Consulting Engineer*, Vol. 40, No. 3, March 1973.
129. World Oil: How Much Is Left? *TIME*, Nov. 19, 1973.
130. Wright, R. A. Lack of Oil Refineries Has U.S. in Bind. *The Akron Beacon Journal*, Dec. 10, 1973.
131. Yellott, J. I. Solar Energy Progress: A World Picture. *Mechanical Engineering*, Vol. 92, No. 7, July 1970.
132. Zarem, A. M., and Erway, D. D. *Introduction to the Utilization of Solar Energy.* New York: McGraw-Hill, 1963.

Index

aeroturbines, 106, 107, 108
agriculture, 2, 30, 85
air pollution, 8, 41, 43, 79, 84
Alaska, 59, 62, 69
Arabian oil, 56, 57, 58, 59, 61
Aswan Dam, 96, 102
automobile, 26, 27, 28, 29, 142

Bab el Mandeb, 103
barrel, 12, 13, 49
batteries, 27, 88, 89
boron, 115, 132
breeder reactor, 120
BTU, definition, 11
butane, 71

calorie, 12
calorific value, 34
chain reaction, 112
Chattanooga shale, 115
city gas, 45, 68
coal
   formation, 33, 34
   production, 35, 37
combustion, 41, 43, 68
control rods, 116
conversion of energy, 10, 131
cost-benefit ratio, 97, 148

dams, 95–100
Dawhat project, 102
depletion allowance, 54
depression power plant, 101
deuterium, 125, 126
direct conversion, 131
distillation, 50, 85
drilling, 53, 70, 80

efficiency, 10, 20, 24, 26, 29, 31, 79, 91, 98, 131
elastic energy, 10
electric
   transmission, 31, 96
   utilities, 18
electron-volts, 15, 125
Energy, definition, 9
engines, 4, 26, 27
enrichment, 112–113
environmental concerns, 8, 39, 55, 63, 64, 65, 71, 79, 93, 100, 114, 122, 123, 127
estuaries, 100, 101
evaporation, 101
exponential form, 12
extraction losses, 52

fission products, 111, 112, 126
foot-pounds, 11

fuel
  cells, 104, 107, 140, 142
  rods, 116
Fundy, Bay of, 100
fusion, 125

gas
  consumption, 69
  diffusion, 112, 113
  formation, 67
  mileage, 27
  pipe lines, 45, 68, 70
  production, 68
  resources, 68, 69, 72
  stimulation, 72
gasification, 45
George's Bank, 60
geothermal gradient, 13, 76
gigawatt, 15
Gross National Product, 6
Gulf Stream, 103

half-life, 112, 127
harnessable water power, 97
heat
  reservoirs, 80
  storage, 90
heating losses, 23
heavy water, 117
heliohydraulic energy, 101
helium, 27, 111, 125
Hoover Dam, 99
horsepower, 13
hothouse effect, 87
hot rocks, 80
Hungary, 25, 78
hydraulic
  energy potential, 98
  turbine, 3, 96
hydroelectric plants, 99
hydrogen, 27, 90, 104, 107, 125, 141, 142
hydrologic cycle, 83

Iceland, 78
Imperial Valley, 79
implosion, 130
incandescent lamps, 20, 21
insolation, 84

intellectual resources, 148
isotopes, 112

joule, 10, 11

kilowatt-hours, 15
kinetic energy, 9

langley, 13
Larderello, 78
laser, 129
Libya, 57, 71, 102
light, 83
lighting efficiency, 21
liquified gas, 71
lithium, 88, 127, 128
lumen, 21

magma, 75
magnetic
  containment, 128
  heating, 129
magnetohydrodynamics, 137–140
Manhattan, 63
megajoule, 11
megawatt, 15
meltdown, 122
meter-kilogram, 11
methane, 45, 67
methanol, 94, 143
microwave, 91
mineable coal, 35

Nadym, 70
natural gas, 67–73
neutron, 116, 125
newton, 11
North Sea, 59
nuclear
  accidents, 114, 122
  energy, 111
  fuel preparation, 112, 114
  reaction, 111–112, 127

ocean thermal gradient, 103
oil
  formation, 51
  imports, 58, 146

oil (*contd.*)
  pipe lines, 62
  production, 57
  resources, 56
  spills, 55
  tankers, 62

photosynthesis, 84
photovoltaic devices, 87, 93, 131
plasma, 129
Plowshare, 71, 82
plutonium, 113, 120
potential energy, 9, 95
power, definition, 13
propane, 71
pumped storage, 100

Quattara project, 102

radioactive
  pollution, 114, 122, 127
  waste, 114, 123
radioactivity, 76, 111
Rance tidal plant, 100
reactor
  accidents, 121
  core, 115
  types, 117–121
refineries, 50
Rocky Mountains, 71

sailing ships, 105, 109
Samotlor, 60
satellite collectors, 90
sea water, 103
Seebeck effect, 133
shale oil, 63–64

shortage, gas, 69
slurry pipe, 40
solar
  cells, 88, 89, 91, 131
  power, 83
status of engineers, 148
steam reservoirs, 78–79
strip mining, 37
sulfur, 39–43, 49, 51, 88
sunshine, 83
supertankers, 62

tar sands, 65
The Geysers, 78
thermal pollution, 84
thermionic devices, 135
tidal energy, 100, 101
tokomac, 128
tritium, 126–127

uranium, 76, 111–117
  reserves, 115

vapor cycle turbine, 80, 104
viscosity, 53
volcanoes, 75, 82

waste, urban, 30–31
water
  heating, 25
  pollution, 40, 55, 65, 79, 123
  power, utilized, 98
  wheels, 2, 95
wave energy, 101
wave energy, 105
women in engineering, 148